Carbon-Based Material for Environmental Protection and Remediation

Edited by Mattia Bartoli, Marco Frediani and Luca Rosi

Published in London, United Kingdom

IntechOpen

Supporting open minds since 2005

Carbon-Based Material for Environmental Protection and Remediation
http://dx.doi.org/10.5772/intechopen.82334
Edited by Mattia Bartoli, Marco Frediani and Luca Rosi

Contributors
Mohamed Deyab, Yıldırım İsmail İsmail Tosun, Şahika Bayazit, Benoit Cagnon, Marius Secula, Mazen Nazal, Rajeev Kumar, Jyoti Chawla, Mohammed Muzibur Rahman, Michela Alfe, Valentina Gargiulo, Brajesh Kumar, Charles Nyanga, Marwa S. El-Azazy, Khalid Al-Saad, Saeed Almeer, Ahmed El-Shafie, Ahmed Issa, Ahmed El-Gendy, Jyoti Chawla

First published in London, United Kingdom, 2020 by IntechOpen
IntechOpen is the global imprint of INTECHOPEN LIMITED, registered in England and Wales, registration number: 11086078, 7th floor, 10 Lower Thames Street, London, EC3R 6AF, United Kingdom
Printed in Croatia

British Library Cataloguing-in-Publication Data
A catalogue record for this book is available from the British Library

Additional hard and PDF copies can be obtained from orders@intechopen.com

Carbon-Based Material for Environmental Protection and Remediation
Edited by Mattia Bartoli, Marco Frediani and Luca Rosi
p. cm.
Print ISBN 978-1-78984-586-0
Online ISBN 978-1-78984-587-7
eBook (PDF) ISBN 978-1-83968-442-5

An electronic version of this book is freely available, thanks to the support of libraries working with Knowledge Unlatched. KU is a collaborative initiative designed to make high quality books Open Access for the public good. More information about the initiative and links to the Open Access version can be found at www.knowledgeunlatched.org

For EU product safety concerns:
IN TECH d.o.o., Prolaz Marije Krucifikse Kozulić 3, 51000 Rijeka, Croatia, info@intechopen.com or visit our website at intechopen.com.

Meet the editors

Dr. Mattia Bartoli began his academic career in the Industrial Chemistry Group at the University of Florence, a multidisciplinary team focused on innovative investigation methods and technological solutions for sustainable waste management and catalyst development. He spent part of his doctoral program as a visiting student at the Institute for Chemical and Fuels from Alternative Resources, developing innovative analytic protocols and biomass conversion procedures. Immediately after obtaining his PhD, Dr. Bartoli moved to the Biorefinery Research Group hosted by the University of Alberta where he developed new materials and technologies. In 2018, he joined the Carbon Group hosted by the Polytechnic University of Turin where he has studied both the production and use of carbon from thermochemical conversion of wastestreams for material science applications.

Marco Frediani is Professor of Industrial Chemistry at the University of Florence, Department of Chemistry "Ugo Schiff," Italy. He obtained his PhD under the supervision of both Dr. Claudio Bianchini, I.C.C.O.M., CNR, Florence, Italy and Prof. Dr. Walter Kaminsky, Insitute of Technical and Macromolecular Chemistry, University of Hamburg, Germany. His scientific interest focuses on catalysis for organic reactions of industrial relevance, polymer chemistry, and reuse of materials at the end of their life cycle by pyrolysis. He has presented at many international conferences, published more than seventy papers in international scientific journals, edited numerous books, and written more than fifteen book chapters. He also holds four patents.

Luca Rosi is Associate Professor of Industrial Chemistry at the "Ugo Schiff" Chemistry Department of the University of Florence. His scientific activity focuses mainly on the field of "homogeneous catalysis," basically on the reactivity of the group VIII metal complexes (Ru, Co, Pd). He also deals with pyrolytic processes that adopt microwaves for the treatment of end-of-life polymeric materials (e.g., tires, post-consumer plastics, solid plastic waste, shredded mixtures, and waste electrical and electronic equipment) and/or biomass in order to convert them to useful products or fuels.

Contents

Preface

Environmental pollution is a global problem that is increasing day by day due to urbanization, industrialization, and human lifestyles [1]. Access to clean air, water, and environments has become extremely difficult for many people. At the same time, environmental pollution has become a challenging task for governments, and mandatory policies for protecting the environment have arisen. Damage to air, water, and natural resources requires remediation. As such, companies and industries have increased accountability to develop sustainable platforms and processes [2, 3].

Nowadays, many technologies are being used to manage and treat environmental pollution and resources depletion. The most widely used technologies are based on carbonaceous materials [4].

Innovative carbon-based solutions can be tuned for many environmental management activities including but not limited to soil improvement, waste management, climate change mitigation, and energy harvesting. Furthermore, rational design of carbon-based technology can help deplete carbon footprint. The spreading of carbon-based technologies represents a game-changing event in the new era of sustainability and environmental preservation.

This book presents an overview of carbon-based technologies and processes, and examines their usefulness and efficiency for environmental preservation and remediation. Topics covered range from pollutants removal to new processes in materials science. This book merges innovative re-thinking of old technologies and problems with the great potential of carbon-based technologies.

Written for interested readers with strong scientific and technological backgrounds, this book will appeal to scientific advisors at private companies, academics, and graduate students. We hope *Carbon-Based Material for Environmental Protection and Remediation* will contribute to focusing the attention of the scientific community on the undeniable threats to our environment.

Mattia Bartoli
Department of Applied Science and Technology,
Polytechnic of Turin,
Turin, Italy

Marco Frediani and Luca Rosi
Department of Chemistry "Ugo Schiff",
University of Florence,
Sesto Fiorentino, Italy

References

[1] Jones O, Preston MR, Fawell J, Mayes W, Cartmell E, Pollard S, et al. In: Pollution: Causes, Effects and Control. Royal Society of Chemistry; 2015

[2] Brandt P, Ernst A, Gralla F, Luederitz C, Lang DJ, Newig J, et al. A review of transdisciplinary research in sustainability science. In: Ecological Economics. Vol. 92. 2013. pp. 1-15

[3] Klewitz J, Hansen EG. Sustainability-oriented innovation of SMEs: A systematic review. In: Journal of Cleaner Production. Vol. 65. 2014. pp. 57-75

[4] Beesley L, Moreno-Jiménez E, Gomez-Eyles JL, Harris E, Robinson B, Sizmur T. A review of biochars' potential role in the remediation, revegetation and restoration of contaminated soils. In: Environmental Pollution. Vol. 159. 2011. pp. 3269-3282

Section 1

Advanced Carbon Materials for Environmental Remediation

Chapter 1

Microwave Carbonation of Thermal Power Plant Flue Gas/ CO_2 by Fly Ash/Coal Char for Soil Remediation and Ground Stabilization

Yıldırım İsmail Tosun

Abstract

In this research, the cementing filler material production by microwave carbonation of flue gas of coal combusting thermal power station of Silopi in Şırnak by fly ash/coal char, Şırnak asphaltite char, in molten alkali salts will be investigated. The optimum carbonation was managed in order to provide an overview of stabilization of foundation grounds. In this study, the effect of microwave energy managed the carbonation by salt slurries with flue gas contents in the reactor. By the slurry character of salts in the furnace during that process, the flue gas of Şırnak thermal power plant, salt type and flue content were investigated for carbonation with weight and slurry performances. In this study, the toxic emitted contents were also determined in salt slurry, regarding the amounts and type of salt contents as sorbent agents. As a result, a significant positive effect of microwave energy on the carbonation products was determined at lower gas flow rate and steam rates. Finally, 23% CO_2 carbonation could be provided. The oil content in flue gas decreased carbonation fractions. The salt slurry content was primarily settled and coal humus char as by-product was also recovered as solid with a 38.7% recovery rate in microwave carbonation in slurry salt with 20% solid/water rate. The strengths of the ground blocks were dispersed to 0.8–1.2 MPa in shear strength and 3.7–9.4 MPa in compression strength. Thus, with the ideal packing, the strength of the mixed cemented blocks produced from these fine fillers and waste mixtures can also reach 11.2 MPa in compression strength and 3.9 MPa in shear strength.

Keywords: carbonation, microwave carbonating, filler, filler blocks, fly ash/coal char, slope stability, Şırnak province, ground stabilization, salt reactor

1. Introduction

CO_2 and HC containing flue gas emissions of thermal power stations pollute the environment and urbanized area, HC emissions of transportation threat hardly and carbonation of flue gas emissions need filtration and carbonization by pressurized water dissolution and reacting by natural alkali lime and magnesia or other alkali sources becomes an industrial advantageous in sequestration. The carbonation sequestration option resulting in green waste solutions or solid fines provided humus or stabilizing filler materials for land. Mg and Ca containing minerals are commonly reacting with CO_2 to form carbonates. Even by evaporation of salty solutions, similar carbonates and sulfates were precipitated at low temperatures such as 40°C. Various types of hot water sources may react with CO_2 to form carbonate regarding salt composition and reaction parameters. Mineral carbonation of CO_2 will also allow using the products in cement industry or as cement material in constructions with low cost. In this study, the supply of CO_2 will be provided by nearby power stations. Other choice will be the purchase of a pure CO_2 originating from flue gas and hydrocarbon dusty mist of coal combusting industrial furnace at a cement plant. For the pilot plant design, a carbonation unit will contain the compressed flue gas tanks so that sequentially compressed CO_2 will be delivered to carbonation reactor in gas phase, temporarily stored underground at site, and conditioned before carbonation. The carbonation reactions to inject the CO_2 at pressurized microwave radiated heating were effective in reaction gaseous phase with salty molten phase at a slightly high pressure and slightly under supercritical temperature. This paper discussed progress on reactor achieved by tests and search for fast reaction methods using exhaust gas containing sulfur and carbon gases at power stations [1–5]. The alkaline sources containing alkali sodium and magnesium salts, under 10–20 bar pressurized CO_2, salt slurry, and additives were searched for microwave carbonation method to enhance mineral reactivity and to analyze the structural changes to identify reaction kinetics and potential impurity and fouling barriers.

Carbonation products of CO_2 gas were Ca and alkali carbonates even metal iron carbonates. Most distinct sequestration is that carbonization outputs a lower energy compound than calcination [6–9]. Calcium and magnesium carbonates commonly occur in nature (i.e., the weathering of rock over geologic time periods). Moreover, the evaporation outputs such as magnesium-based minerals are dissolved by hot waters and then crystallized at 30–40°C coming out as evaporates on earth. The evaporate carbonates are resistively stable and thus do not re-evolve CO_2 into the atmosphere as an issue. However, settling carbonation evaporates are crystallized very slow under warm temperatures and even saturated in effluent warm waters [10–12].

Natural gas, internal engine combusting and coal fired combusting systems account for almost 80% of the total of world carbon emissions today. There is an important need for carbonation in eliminate carbon gases emission to nature, ease of use and storage, existing filler structure, and most low cost rather than amine absorption. Forty percent of global electricity is generated in fossil fuel power plants per annum, with emissions of about 23% of global energy-related CO_2 pollutes (5.5 billion metric tons) of about 14.7 Gt in 2015 [11, 12]. Over a quarter of the electricity demand of Turkey is supplied by coal-fired power plants, with emissions of about 4 million metric tons of CO_2 as pollutant, among about 23% energy-related polluting gas emissions [13, 14].

Therefore, sequestration with effective CO_2 carbonation method is one of the critical choices in addressing global warming and air pollution. It is improving the efficiency of fuel utilization and curing the environment. The renewable energy sources will certainly play a very important role in reducing CO_2 emissions [15]. Those carbonization and amine absorption methods alone cannot address the greenhouse emission issue mainly because world energy consumption will increase significantly as the living standard improves in many parts of the world. The coal combusting boiler types of power plants and internal vehicle engines still emit over 5% of the carbon dioxide, 1% of the sulfur dioxide, and less than 1% of the nitrous oxide emitted by a coal-fired plant. Similarly, injection of compressed gas to cold back injection to geothermal fields sequestrated less of the carbon dioxide as shown in **Figure 1** [16].

The method for storing CO_2 in deep underground geological formations need adequate porosity and thickness for storage capacity, and permeability for gas injection that are critical as shown in **Figure 2**. The storage formation should be capped by

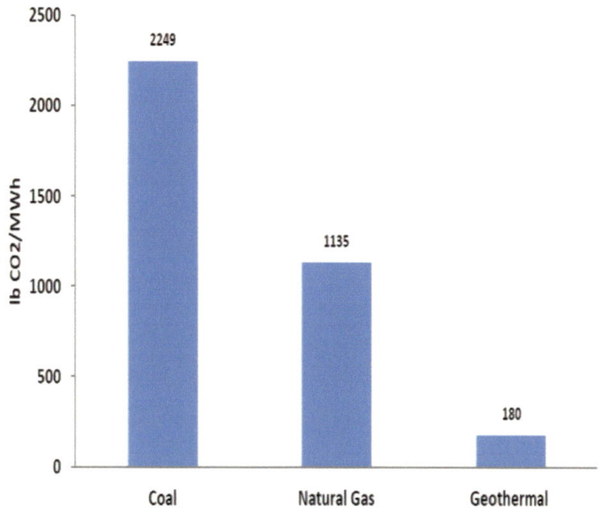

Figure 1.
Comparison of CO_2 emissions by coal natural gas and geothermal source.

extensive confining units such as shale, salt caves or anhydrite beds to ensure that CO_2 does not escape into overlying, shallower rock units and ultimately to the surface [17].

1.1 CO_2 capture and separation from flue gas thermal power plants

The abatement of greenhouse gases is becoming increasingly important. In the last decades, the government issued a White Paper on Energy outlining the national strategies for abating greenhouse gases and filtration systems use in thermal power plants. One of the key options highlighted was the eliminating fly ash and hazardous toxic gas emissions cut and even CO_2 from stack emissions and subsequent storage in geological reservoirs (carbon capture and storage can be retrofitted easily onto the tail end of power-plant flue gas streams without requiring complicated integration in Turkey) [18, 19].

Several technologies have been proposed to capture CO_2 from power-plant flue gas including absorption, adsorption, cryogenic distillation, and membrane gas separation. The technology examined in this article is polymer-based membrane gas separation. The membranes have been used commercially to refine natural gas and biogas for CO_2 [20, 21].

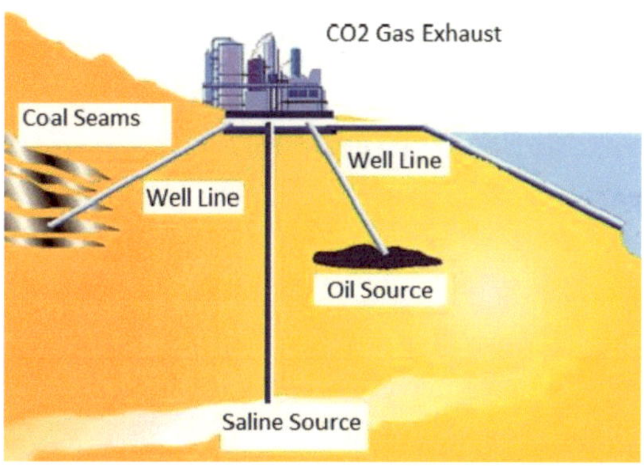

Figure 2.
General chemical carbonation or sequestration path.

1.1.1 Amine absorption

From Fick's law (Eq. (1)), increasing the selectivity of CO_2/N_2 increases the mole fraction of CO_2 in the permeate (y_i) and decreases the mole fraction of CO_2 in the retentate (x_i). Consequently, the driving force across the membrane is also reduced. To obtain the same amount of CO_2 recovered (J_i), the membrane area (A_m) increases and thus the capital costs for membranes also increase. However, as the CO_2 purity of permeate stream (y_i) increases, the flow rate of the permeate decreases, resulting in a smaller post-separation compressor. It is the balance between the cost savings generated by the post-separation compressor and the increase in membrane costs that influences the cost trends [22–24].

1.1.2 Membrane separation

The effective force across a gas-separation membrane is the pressure differential (ΔP) between the feed side and the permeate net, and CO_2 emissions exceeded 550 million tons in 2013. Over 45% were from post-combustion coal-pulverized power plants, and because of this, it is likely that these will be the initial focus of CCS [25–27].

As given below with Fick's law (Eq. (1)), increasing the content of CO_2/flue gas increased the mole fraction of CO_2 in the permeate (y_i) and decreased the mole fraction of CO_2 in the retentate (x_i). Finally, the efficiency of passage across the membrane was decreased. In order to receive high of CO_2 passage (J_i), the membrane area (A_m) increased and thus the capital costs required membrane also increased. Meanwhile, the purity of CO_2 stream (y_i) increased, and the flow rate of the permeate decreased the cost savings generated by the post-separation compressor and the increased the membrane costs that influenced the separation trends [25–29].

Fick's Law:

$$Ji=Yi/Am=-Xi/Am=-Di\ dCi/dt=-Di\ dPi/dt \quad (1)$$

where mass diffusion rate Ji, mass rate of i per area A at diffusion rate constant Di, at pressure or concentrate of permeate change per thickness t.

As the membrane used at atmospheric pressure, compressed feed gas to a high pressure was feasible for CO_2 capture in membrane technology. In our previous work, we have shown that the cost is high because of the high capital costs associated with compressors needed to compress the low-pressure flue gas and the low CO_2 purity product stream. As the feed gas is compressed to 15–20 bar, the recovery of CO_2 was roughly managed [30]. Those case studies showed that the cost for CO_2 capture was at least 30% higher than for CO_2 recovered using amine chemical absorption.

2. Production of reactive coal char and biomass char/char carbon in Şırnak

The energy production by combusting asphaltite, a type of coal, by local alkali rocks such as limestone is advantageous for the development of the South-East Anatolian region and also the industrial construction and diversification [1–5]. As given in **Table 1**, the thermal power station in the region combusting Şırnak asphaltite by addition of 15% limestone şn fluidized bed combustion system emitting much clean exhaust gas out regarding mid power output [31–33].

Flexible and regional targets for a mobile solid waste incineration from an environmental and economic perspective were the following:

- The mobile plant where the waste sorting process is performed can be processed to acquire secondary materials
- Biological treatment of biomass and conversion to compost, which is a market value or energy production by producing methane gas by anaerobic treatment
- Recycling and reducing the amount of waste storage following thermal systems, making it inert and obtaining energy as given in **Table 2**
- Regular land filling and use of landfills for land reclamation and at least the reduction of pollution in the Şırnak Province

Biochar was beneficial in soils around the land as a result of vegetation fires and soil management practices. Intensive study of biochar-rich dark earths in the Amazon (terra preta) has led to a wider appreciation of biochar's unique properties as a soil fertilizer [34–36].

The carbon in biochar resists degradation and can hold carbon in soils for hundreds to thousands of years. Biochar is produced through pyrolysis or torrefaction processes that heat biomass in the absence (or under reduction) of oxygen.

In addition to using soil fertilizers, sustainable biochar dissociation can produce black liquor, oil by-products that can be used as fuel, providing thermal energy. When the biochar is buried in the ground as a soil improver, the natural system may be "carbon negative" [37–39].

Biochar and bioenergy co-production can help combat global climate change by displacing fossil fuel use and by sequestering carbon in stable soil carbon pools. It may also reduce emissions of nitrous oxide **Figure 2**.

Mobile incineration and waste management in Şırnak included energy technologies from recycling, composting and baling waste in order to distribute to long distances. For this, collected waste was classified and iron scrap was send to Iskenderun steel plant. The debris parts were used in road pavements. The combustible waste products controlled organics and plastics. The mayor was aware that the organic products to be obtained from urbanized site should be processed by the sorbent in incineration. These markets are also likely to be sensitive to the quality and quantity of the supply. The distribution of solid wastes in Şırnak Province is shown in **Tables 3** and **4**.

2.1 Pellet char production

Mobile waste management was flexible in terms of incinerating small amounts of waste, as operation, even if it needed to adapt to existing waste type and environmental conditions in the Southeastern Anatolia. Mobile incineration provided the flexible solid waste management of flexible transfer of small amounts of waste to

Thermal power plant size (net)	415 MW
Boiler type	Supercritical
Coal type	Asphaltite
Thermal efficiency (LHV)	38%
Temperature	500
Pressure	85 bar

Table 1.
Processing conditions and composition of a typical supercritical bituminous power-plant flue gas in Şırnak.

Total	Landfill	Incineration	Energy recovery	Backfilling	Recycling	
(million tons)			(%)			
EU-28	2319.5	47.4	1.5	4.7	10.2	36.2
Belgium	42.8	8.2	4.3	13.6	0.0	73.9
Bulgaria	175.7	97.9	0.0	0.1	0.0	2.0
Czech Republic	19.9	17.3	0.4	5.1	29.1	48.1
Denmark	17.7	21.7	0.0	20.7	0.0	57.6
Germany	370.7	19.2	2.3	10.5	25.3	42.7
Estonia	20.7	65.6	0.0	2.5	11.9	20.0
Ireland	10.0	42.6	0.1	7.2	37.4	12.7
Greece	67.1	88.4	0.0	0.2	8.1	3.2
Spain	103.4	47.9	0.0	3.4	12.6	36.1
France	299.7	29.3	2.0	4.5	10.7	53.6
Croatia	3.5	51.1	0.0	1.4	2.0	45.5
Italy	129.2	16.0	5.2	1.6	0.2	76.9
Cyprus	1.8	58.9	0.0	1.7	25.9	13.5
Latvia	1.9	34.8	0.0	8.7	0.9	55.5
Lithuania	4.5	67.6	0.1	4.1	2.5	25.8
Luxembourg	8.5	38.3	0.0	2.5	16.0	43.3
Hungary	13.7	39.4	0.7	8.9	3.7	47.3
Malta	1.6	28.9	0.4	0.0	37.5	33.3
The Netherlands	130.6	45.4	1.0	7.9	0.0	45.7
Austria	53.9	38.6	0.2	6.5	20.1	34.7
Poland	182.4	24.9	0.4	2.7	21.5	50.5
Portugal	9.9	31.8	10.0	3.1	0.0	55.0
Romania	172.2	94.4	0.0	1.3	0.6	3.7
Slovenia	5.4	9.2	0.6	4.9	33.5	51.8
Slovakia	7.1	53.8	0.8	4.4	0.0	40.9
Finland	93.3	80.9	0.5	4.8	0.0	13.8
Sweden	163.3	84.4	0.1	4.7	1.6	9.3
The United Kingdom	209.0	41.5	3.6	0.9	10.4	43.6
Iceland	0.5	30.7	0.0	2.7	0.6	66.0
Norway	11.7	17.9	0.5	35.8	5.3	40.5
Montenegro	1.0	98.8	0.0	0.1	0.0	1.0
Form. Yug. Rep. of Macedonia	1.5	98.7	1.3	0.0	0.0	0.0
Albania	1.2	74.8	3.1	0.5	0.0	21.6
Serbia	49.4	97.3	0.0	0.1	0.0	2.6
Turkey	79.3	70.2	0.0	0.7	:	29.0

Table 2.
Distribution of urban wastes conversion and energy generation by European countries, Eurostat waste management Statistics 2016 [40].

Waste type	Theoretical amounts, tons per annum		
	Regular	%	Heat value, kcal/kg
Agricultural waste	13,000	14.5	2100
Cardboard	12,000	12	2100
Poultry	9000	9	1100
Sewage	32,000	33	1100
Dairy	1000	1.2	1700
Forest	32,000	33	3100

Table 3.
Biomass waste type of Şırnak Province and distribution in quality.

Biomass char	C, %	Ash, %	Moisture, %	S, %	P, %	Na + K, %	Mg,%	Ca
Agricultural waste char	13	3.4	69	0.1	1.1	1.1	1.1	1.1
Cardboard char	12	0.5	44	0.1	0.1	0.1	0.1	0.1
Poultry char	9	8.9	65	0.9	5.1	3.1	2.1	3.1
Sewage char	2–3	77–87	11–19	0.3	2.1	2.1	2.1	2.1
Dairy char	1	4–5	88	0.2	4.1	4.1	1.1	4.1
Forest waste char	32	1	55	1.3	1.1	1.1	2.1	1.1

Table 4.
Chemical carbonation quality of biomass waste char of Şırnak Province and distribution in quality.

direct treatment systems as much hard environmental conditions [41]. The drying and torrefaction were packing and pelleting of biomass wastes managed for Biomass/Waste Drying System Pelleting System/Cooling /Sieving/Bagging System, Wood Char Pelleting (**Figure 1**).

2.2 Mobile coal char pelletization in Şırnak

In the Project, the scope of this study is 5 MW with regional biomass waste and Şırnak asphaltite primary energy source to evaluate Şırnak's biomass for electricity generation. The common burning of biomass resources within the special firearms is designed and proposed by providing legal and institutional, economic and environmental impact assessment. However, the use of Şırnak biomass energy source is to develop solutions against technological coal burning pollution. Biomass wastes for most of Anatolia, Southeastern Anatolia Region and Şırnak Province in Eastern Anatolia and Southeastern Anatolia region, which mostly contains high elevation figures in Turkey, and solid fuels are consumed. In addition, energy production is realized with thermal power plants especially Şırnak asphaltite and Afşin Elbistan lignite [42, 43] in the analysis, as the coal slime Sırnak asphaltite slime samples was used and the char as shown in **Figure 4**, the reduction of the coal samples was shown in semi pyrolysed fractions. The chemical analysis temperature was continuously weighed and the combustion analysis was carried out in the bath oven. The test results are shown in **Figure 5** for biomass pellets and coal sample. As shown in **Figure 6**, the effect of addition is determined in combustion experiments, the lime on desulfurization and emission is hydrated, and the reactor temperature was 500°C and only 10% MgO char pellet at weight rate. The temperature varied

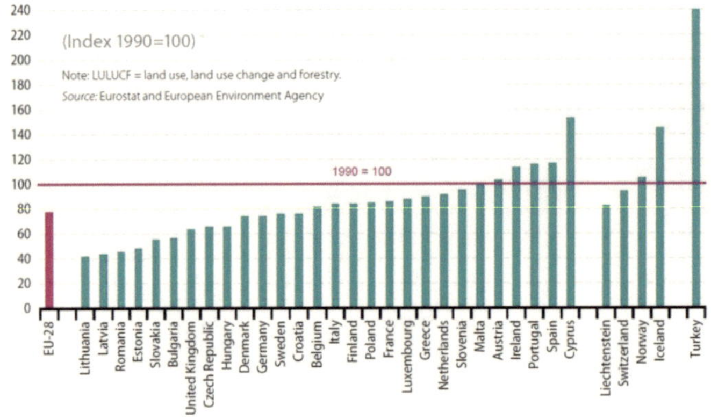

Figure 3.
Green gas CO_2 emission to environment in EU28, Eurostat 2016 [40].

to 550°C and solid waste mixture samples were analyzed for sulfur content on semi pyrolysis char samples. Experimental results are shown in **Figure 6**. For the production of pyrolysis oil from a mixture of waste at weight rate, 50% animal manure and 50% human manure at three production stages were put in the kiln at 1 h, 8 h, 14 h and at total 50 h pyrolysis completed. The pyrolysis oil product was in total over 9 kg/100 kg. The production was executed using a feed rate of 150 kg/h, with average pyrolysis temperature of 500°C (**Figure 4**).

The mass, energy and carbon balance for the pyrolysis of animal manure and human manure compared by Şırnak asphaltite in char production was presented in **Table 3**. The pyrolysis oil and char yield (carbon basis) were significantly lower for the manure compared to coal. That was for the reason of higher moisture content of the type manure wastes so was over than 80%, far optimal processing, which often resulted in greatly dişturb the black liquor oil production. Especially the char yield was higher for the manure pyrolysis even providing phosphate ash, which could very likely be advantageous in emission control sorbent production. In a full-scale pyrolysis plant, the energy value from the char was not lost, but will be recovered via decantation systems. The high yield of water phase for the manure waste types was not a suitable indirect char output as the result of the weight of char next to high moisture content evaporated hardly from the feedstock. Indirectly, the phosphate content of manure types also increased high ash content of the char due to phosphate ash. Ash components such as phosphor and potassium were known to enhance the sorption gaseous components as the desulfurizing components. The gas yield was not higher for the manure pyrolysis case, and this was about the lower pyrolysis temperature, in combination with a lower mass balance output received as given in **Table 5**. The analysis showed that soot formed

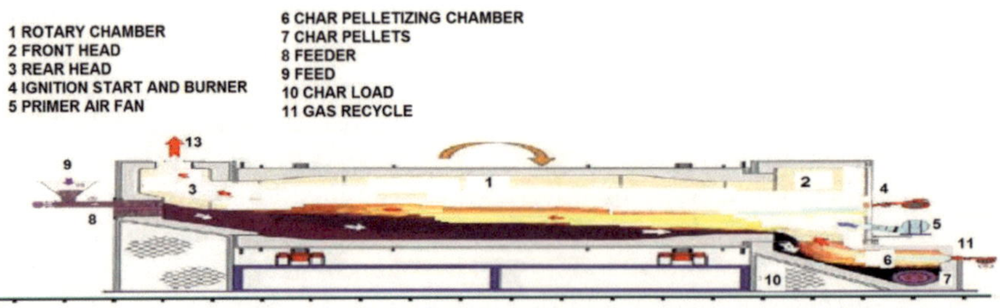

Figure 4.
Biomass/waste drying system pelleting system/cooling /sieving/bagging system, wood char pelleting.

from the light hydrocarbons within the reactor or unconverted char particles formed during the conversion of the pyrolysis oil. Animal manure and human manure could be converted into pyrolysis oil at laboratory scale. The pyrolysis oils were converted into syngas in a fluid bed flow gasifier, which was operated at a capacity of 2 kg/h pyrolysis oil input. Temperatures around 950°C were obtained, yielding a methane-rich syngas product with volume fractions of 26% CO, 10% H_2 and 10% CH_4, 13% CO_2 on dry, and N_2 free basis for both pyrolysis oils. Animal and human manure-derived pyrolysis oil was successfully converted into methane-rich syngas by Şırnak asphaltite.

In the pyrolysis experiments with addition of hydrated lime, reactor temperature changed between 400°C and 650°C and asphaltite samples mixed only by %10 lime. Products received from pyrolysis of coal specimens were subjected to analysis for sulfur holdup managed effectively (**Tables 5** and **6**).

The reactive contents of produced semi carbon char pellets are given in **Table 6**.

With the mobile pyrolysis system in Şırnak Province, the urbanization organic waste transformed into energy and fertilizer. For direct heating systems used in Şırnak, thermal insulation and coal boilers feed reduced 4–5% by weight of fuel. The boiler bottom ash could be used as filler material in cement and lime plants close to the locality, 10% by weight potentially. The fly ash of Silopi thermal coal power plant was used as sorbent in fluidized combusting system in order to reduce hazardous gas pollution at stack of plants.

Figure 5.
Reactive biochar picture for carbonation and carbon use.

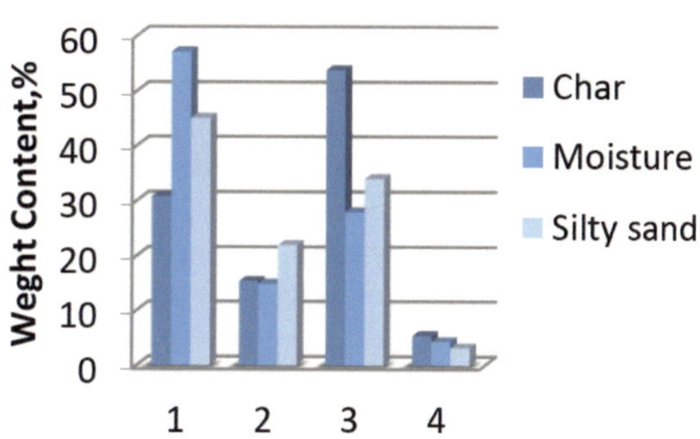

Figure 6.
The component distribution of reactive char of different biomass waste sources for Şırnak biomass and asphaltite coal. (1) Coal Slime, (2) Municipal Sewage Sludge, (3) Forest Biomass Waste, and (4) Mixture semi Char.

	Animal manure	Human manure	Asphaltite
Pyrolysis oil			
Mass (g kg^{-1})	84.1	55.7	143
Energy (J J^{-1})	0.588	0.553	0.47
Carbon (mol mol^{-1})	0.582	0.420	0.382
Water phase			
Mass (g kg^{-1})	457	526	25
Gas			
Mass (g kg^{-1})	154	132	334
Energy (J J^{-1})	0.068	0.049	0.068
Carbon (mol mol^{-1})	0.120	0.13	0.120
Char			
Mass (g kg^{-1})	97	87	140
Energy (J J^{-1})	0.166	0.141	0.266
Carbon (mol mol^{-1})	0.073	0.070	0.163
Ash (g kg^{-1})	54	94	360
Total			
Mass (g kg^{-1})	999	977	999
Energy (J J^{-1})	0.924	0.960	0.924
Carbon (mol mol^{-1})	0.979	0.935	0.979

Table 5.
Mass, energy and carbon balances for the products in char production for carbonation/carbon use of flue gas of thermal power plants.

Şırnak asphaltite + sewage content, %	P, g	C, g	N, g	K, g	Na + K, g	Ca, g	Mg, g
A. 82 + 18	323.3	1123.3	23.3	23.3	323.3	323.3	323.3
B. 70 + 30	744.5	2214.5	445	44.5	744.5	744.5	744.5
C. 50 + 50	2122.2	4442.2	2220	162.2	2122.2	2122.2	2122.2

Table 6.
The humus chemical contents for carbonation activity of produced semi char pellets from sewage sludge and Şırnak asphaltite mixture chars used in microwave carbonation tests.

3. CO_2 sequestration and carbonization

Sequestration deposits of CO_2 require salt or geothermal space underground for compressed CO_2 to allow injection. The compressed CO_2 reaches a phase known as 'supercritical.' This state is achieved by exposing the CO_2 to temperatures over 31.1°C and pressure over 73.9 bars. The sequestration density of CO_2 will be managed by geological deposition depth, until about 800 m or over a dense supercritical state [15, 16].

The mineral carbonation, a process of converting CO_2 into stable sequestration, was studied extensively to capture and store CO_2. However, most of the Ca and Mg, forsterite sequestration studies were investigated at lab scale. Preliminary and pilot scale studies for accelerated mineral carbonation (AMC) were conducted at one of the largest coal-fired power plants (2120 MW) in the USA by reacting flue gas with fly ash particles in a fluidized bed reactor. In the preliminary experiments, flue gas CO_2 and SO_2 concentrations decreased from 13.0 to 9.6% and from 107.8 to 15.1 ppm, respectively, during the first 2 min of reaction. The flue gas treated by fly ash particles, even mineralization hold high mercury (Hg) concentration of 0.22 mg/kg in flue gas (**Figure 3**) [15, 16].

Fly ash utilization and using CO_2 as feed material in mineral carbonation produce various environmentally benign products. However, many challenges to any solution include technical feasibility, economic viability, environmental soundness and long-term sustainability for mineralization of CO_2 [13–15].

4. Method of microwave carbonation

Mineral carbonation reactions are known to geologists and occur spontaneously on geological time scales [43–64]. For example, the reaction of CO_2 with common mineral silicates to form carbonates like calcite and magnesite or calcite is exothermic and thermodynamically favored [64–78]. The design for carbonation microwave furnaces is shown in **Figure 7**.

The reactions given below transformed as that the CO_2 gas to Ca carbonates by Reaction 1 has the potential to convert naturally occurring silicate minerals to hydrated and carbonated minerals of silica and silicates. The process followed natural chemical formations such as altering of rocks to form carbonates over million year time periods. Reaction 2 illustrates the transformation of the common silicate mineral serpentine, $Mg_3Si_2O_5(OH)_4$, and CO_2 into magnesite, $MgCO_3$, silica and hydrate. At theoretical case, a unit of serpentinite can dispose of approximately one-half of CO_2. Reaction 3 illustrated the conversion of forsterite and was the final common silicate mineral olivine. A unit of olivine can dispose of approximately two-thirds of CO_2. Again, the reaction is exothermic and releases 90 kJ/mole of CO_2.

$$(Mg, Ca)_x Si_y O_x + 2_y + zH_2O + xCO_2 \rightarrow x(Mg, Ca)CO_3 + ySiO_2 + zH_2O \quad (2)$$

$$1/3Mg_3Si_2O_5(OH)_4 + CO_2 \rightarrow MgCO_3 + 2/3SiO_2 + 2/3H_2O + 64\ kJ/mole \quad (3)$$

$$1/2\ Mg_2SiO_4 + CO_2 \rightarrow MgCO_3 + 1/2SiO_2 + 90\ kJ/mole \quad (4)$$

An illustrated in the process shown in **Figure 8** of the searched the carbonization process as presented in **Figure 3**. CO_2 from one or more power plants is transported to a carbonation reactor, combined with fly ash slurry treated at certain appropriate reaction conditions until the desired degree of carbonation was reached. The resulted solution chemistry yielded olivine conversions of 90% in 24 h and 83% within 6 h. The study shows that further modifications of the same basic reaction can achieve 65% conversion in 1 h and 83% conversion in 3 h [16]. A recent literature review indicated that weak carbonic acid treatments had also been suggested for Mg extraction in the prior literature [17]. Carbonation tests at ARC resulted in heat pretreated serpentinite conversion up to 83% conversion in 30 min lower than 115 bars [18]. By increasing sodium bicarbonate concentration, the carbonation reaction of serpentinite can reach 62% conversion under 50 bars.

The products of carbonation, which were slurries of hydrated and carbonated minerals and lean gas CO_2 in aqueous, were followed by decantation separation. The exhaust CO_2 is recycled. The finer solid matter was transferred as filler for

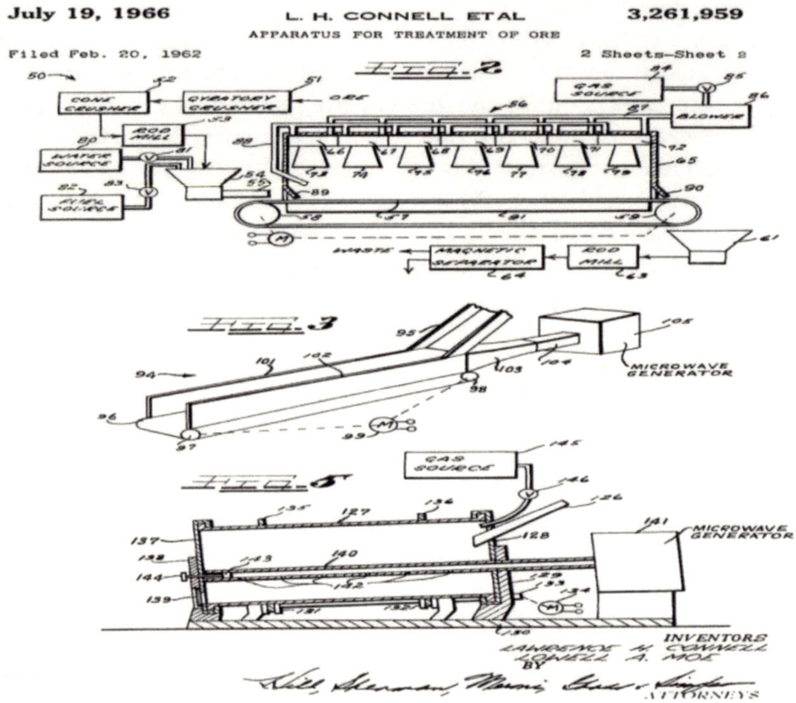

Figure 7.
Types of CO_2 microwave carbonation furnace for sequestration path by fly ash.

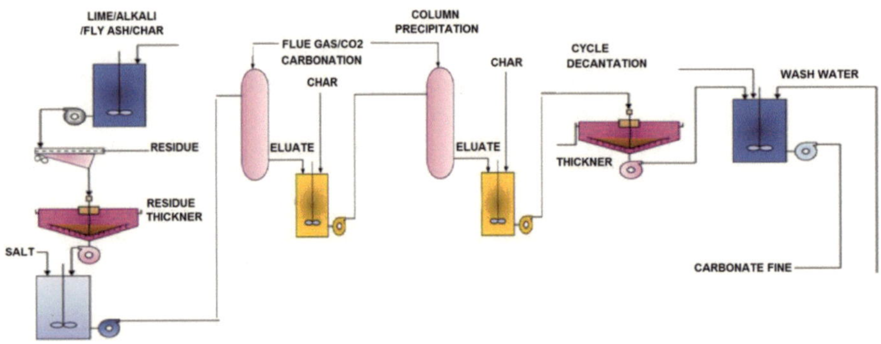

Figure 8.
CO_2 carbonation or sequestration path by fly ash.

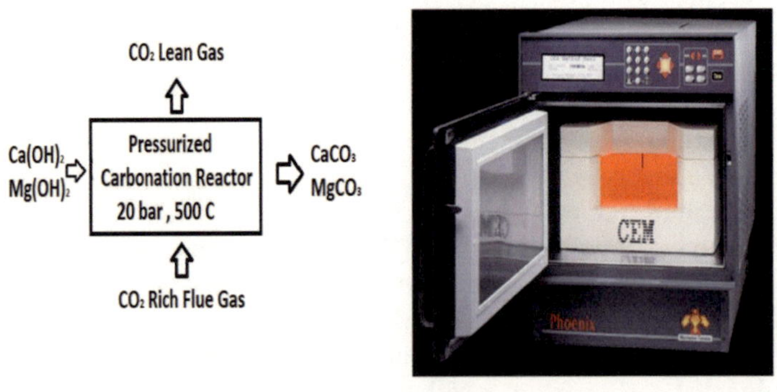

Figure 9.
(a) Microwave laboratory steel jar CO_2 carbonation by fly ash/coal char and (b) microwave ash laboratory furnace.

construction works or the carbonated materials were returned to the caving in mine site. Almost calcium and magnesium oxide (MgO) content in the magnesium silicate minerals were of 40% and 60–70% reactive efficiency for the carbonation regarding mass balance of Eq. 1. A 100 MW power plant in Şırnak, generating approximately 200 tons/day of CO_2, would require just over 100 tons/day of calcium and magnesium containing fly ash. Several fly ash types in Turkey contain sufficient calcium and magnesium oxide quantity in silicate mineral to provide raw materials for the mineral carbonation (**Figure 9a** and **b**).

5. Experimentation method of microwave carbonation

In this research, representative specimens of the different types of Turkish fly ash/coal char sources in Şırnak were classified to calcium content and bicarbonate by chemical analysis. Gas samples of 10–20 kg tubes from Silopi coal power stations were used. Chemical analyses of salt mixtures used in the experiments are given in **Table 7**.

Chemical reaction of molten salt sources is illustrated in **Figure 6**. Carbonation salt mixtures contain reactive 25, 35 and 45% hydrated lime and rest reactive caustic, respectively. Salt mixtures contained microwave radiation emitter 2.2, and 5.7% serpentinite and also non reactive, respectively. The crystal waters and moisture of salt hydrates dissociate under microwave radiation and salt mixtures resulted in various sulfate compounds. Substantial fractions of these compounds in carbonates could be stuck during clay dissociation to pore leaks. Fifty–sixty percent of arsenic, lead, manganese, mercury and selenium could be removed by solid salt oxides in transforming during cooling.

Screen analysis of Turkish lignite fly ash samples was done by standard Tyler screens and particle size distributions and normal distributions of lignite samples were tested. Specific surface area fly ash/coal char samples were about 1.76–4.2 m^2/g determined by BET surface analyzer and highly sufficient in order to react with gaseous CO_2. About 80% of weights of samples were 0.1 mm. Ash particles in lignite samples were mainly distributed below 40 and 100 μm size fractions. Main reactive silicate structure is widely distributed and pore structures are associated with ash minerals. Coarse alkali oxides are also seen in the picture shown in **Figure 10**.

Salt mixture composition, g/kg	Ca1	Ca2	Ca3	Mg1	Mg2
Ca^{2+}	550	422	220	43	26
Fe^{2+}	7.12	3.69	1.1	0.2	0.5
K^+	63	79	59	102	130
Mg^{2+}	10.6	6.7	3.0	489	310
Mn^{2+}	0.6	0.02	—	0.01	0.01
Na^+	15	18	6	16	9
F^-	0.35	0.51	0.55	—	—
Cl^-	19	122	250	27,440	114,300
NO_3^-	1.34	0.9	0.92	—	—
SO_4^{2-}	13.43	14.5	14	11.34	35.47

Table 7.
Chemical analyses of salt mixtures used in the microwave carbonation experiments.

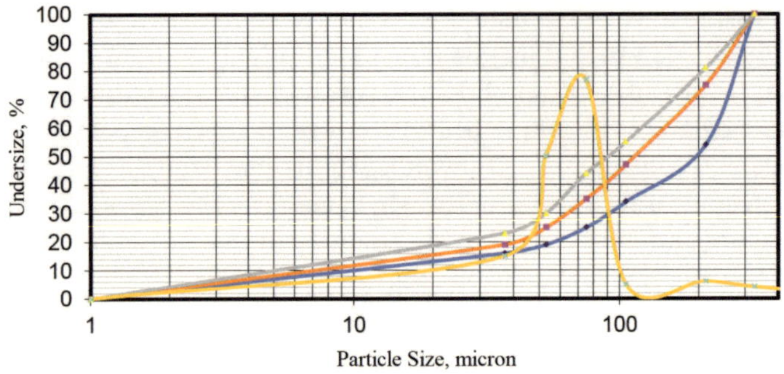

Figure 10.
Particle size distributions and normal size distributions of fly ash/coal char samples tested.

6. Results and discussions

The major technical challenge was hindering the use of carbonation method for CO_2 due to low reaction rate. The reactivity of rock was extremely slow. The priority was given to improving faster reaction pathways. The optimized process should be economical. Although the carbonation reactions were exothermic, it is generally very low-grade heat with the long reaction time and demanding uncontaminated reaction conditions (**Figure 11**).

The environmental impact from mining, mineralization and carbonation processes must be considered in carbonation. We succeeded in achieving shortened carbonation reaction times employing fly ash/coal char containing lime and calcium magnesium silicates such as gehlenite and mehlenite. Reaction took 4 h to reach 40–50% completion of carbonation with fly ash/coal char. The reaction required microwave temperatures of 450–550°C, pressures of 10–20 bar (**Figure 12**), and mineral particles in the −100 μm size range.

Because the high pressure requirement of the carbonation reaction will certainly lead to high process costs, the team is modifying solution chemistry to allow reaction to proceed at a lower pressure and temperature. The research showed that the concentration of sud caustic in the solution was critical to the reaction rate. The high CO_2 pressure will lead to increased CO_2 absorption in the solution and thus enhance the liquid concentration. Adding bicarbonate such as sodium bicarbonate in the solution will significantly increase the liquid concentration even at a relatively lower CO_2 pressure (**Figure 13**).

In the carbonation experiments with addition of fly ash/coal char, reactor temperature changed between 400°C and 650°C and Silopi fly ash/coal char samples mixed only by 10% rate. Products received from carbonation of ash specimens were subjected to analysis for gas holdup determination. Test results of carbonation by fly ash/coal char and also salt mixtures are seen in **Figure 10**.

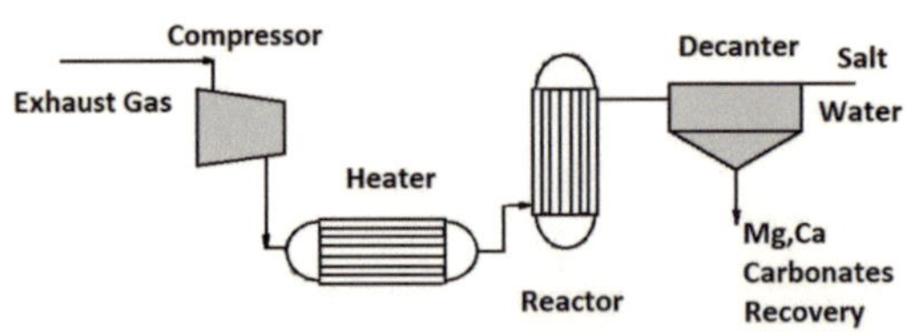

Figure 11.
Reaction diagram of molten salt mixtures and fly ash/coal char used in carbonation process.

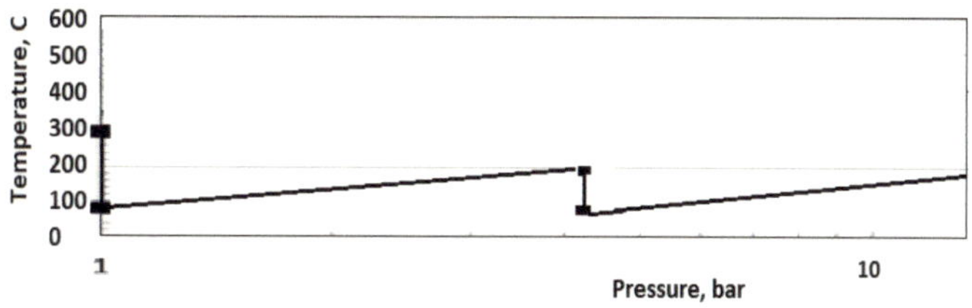

Figure 12.
Pressure-temperature steps for microwave carbonation process.

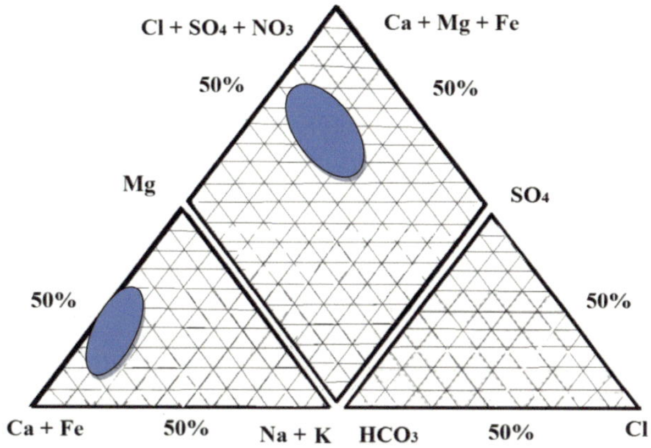

Figure 13.
Piper diagram of salt mixtures used in microwave carbonation process.

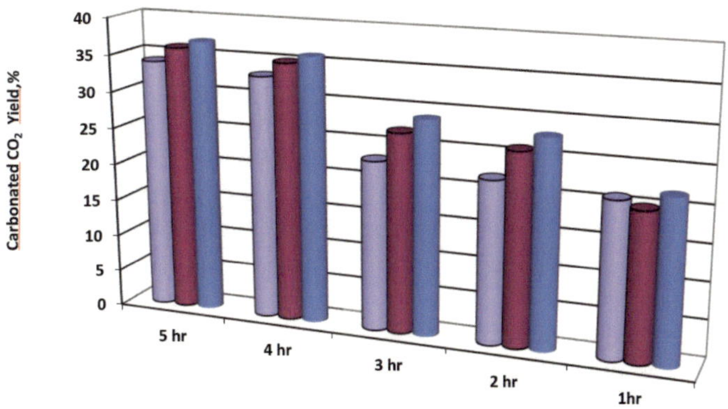

Figure 14.
Effect of carbonation time for microwave carbonation over conversion yield rates in salt mixtures used.

From the point of view of carbonation experimentation, Silopi flue gas carbonation value was significant. The quantity in the carbonation chambers for flue gas samples was determined for different source evaluation and for reducing the effect of ash content of coal samples in order to optimize carbonation rates of fly ash/coal char samples. As given in **Figure 14**, gas conversion yield for salt mixtures and alkali caustic acts were distinctly improved.

7. The column reactor of microwave carbonation process

In the carbonation experiments with addition of fly ash/coal char, reactor temperature changed between 400°C and 650°C and Silopi fly ash/coal char samples mixed only by 10% rate. Products received from carbonation of ash specimens were subjected to analysis for gas holdup determination. Test results of carbonation by fly ash/coal char and also salt mixtures are seen in **Figure 15**.

From the point of view of carbonation experimentation, Silopi carbonation value was significant. The quantity in the carbonation chambers for Ca 2 samples were determined for different source evaluation and for reducing the effect of ash content of coal samples in order to optimize carbonation rates of fly ash/coal char samples. As given in **Figure 15**, gas conversion yield for salt mixtures and alkali caustic acts were distinctly improved.

The tests with the microwave carbonation by fly ash packed bed under the high-pressure feed column process were carried out. As shown in **Figure 15**, for column permeate systems, the gases were recycling at three steps. As shown in **Figure 16**, the melted salt types affected greatly carbonation conversion under microwave radiation as giving the products of 36 gr sodium and Ca carbonates by fly ash added at weight rate of 10%.

At the design stage of the unit, it was decided that the overall capital cost of the sequential column system was much higher than that for the high-pressure feed packed bed column method (**Figure 17**). The column permeates on the recycling gas stages were at the largest cost, accounting for over 10% of the total equipment costs. In comparison, for the high-pressure feed process, the cycling time cost accounts for less than 70% of the total equipment costs. For the microwave carbonation by coal char tests under column gas flow conditions, the results were shown in **Figure 18**. As shown in figure, the melted salt types affected greatly carbonation conversion under microwave radiation giving the products of 64 gr sodium and Ca carbonates by coal char, Şırnak asphaltite char type A, added at weight rate of 10%.

Over 10 bar pressure, the weight percent increases in salt mixtures for pressurized column processes. First, without the feed-gas compression, the overall total carbonate conversion decreased. Second, the column pressure difference in caustic permeate steps was less than 2 bar that for the high-pressure feed process at 12 bar. In this analysis, the amount of CO_2 recovered per second (J_i) was fixed at 85–90% of the flue gas or 2.49 kmol/s. In addition, P_i^* and δ were constant. Because ΔP is reduced for the column rough stage, Fick's law (Eq. (1)) showed that the required membrane area (A_m) should increase to maintain a fixed value for J_i. Hence, the cost of the microwave carbonation three step stream column unit increased under pressurized cycling conditions.

The second-largest conversion step was the post-cycling conversion, also shown in **Figure 18**. The rough column unit and scavenger CO_2 column together accounted

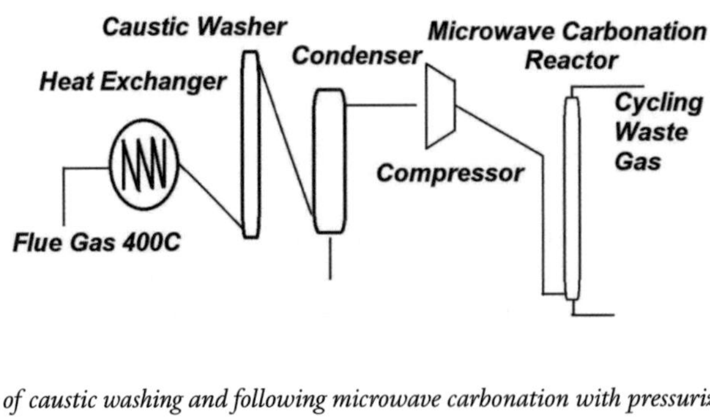

Figure 15.
Simplified diagram of caustic washing and following microwave carbonation with pressurized conditions.

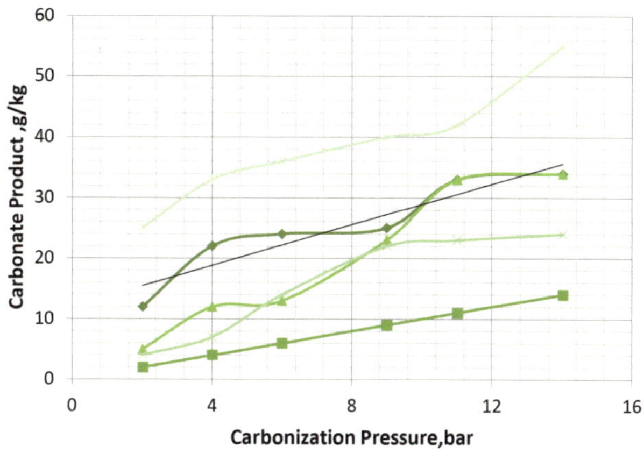

Figure 16.
Percentage melted salt carbonate product by column pressurized feed in microwave carbonation by Şırnak asphaltite char at 10 wt% systems under high pressure (HP) feed conditions.

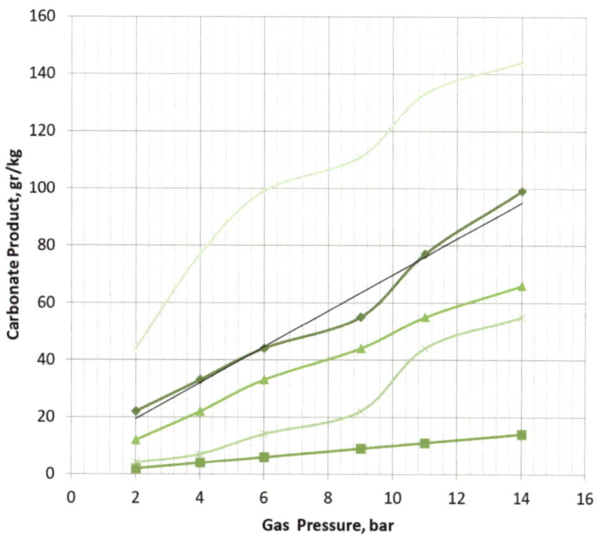

Figure 17.
Percentage melted salt carbonate product by column pressurized feed in microwave carbonation by Şırnak asphaltite char at 10 wt% systems under high pressure (HP) feed conditions.

for approximately 75% of the total salt melting system. **Figure 19** illustrates that the two most expensive items are replacing the membranes and the energy required for compression. The results from **Figures 17** and **18** indicate that carbonate weight could be achieved by the following:

The effect of CO_2 content was ascertained as depended on flue gas type. The CO_2 content will influence the rate at which CO_2 is carbonated from the feed gas. For a fixed flux of CO_2 across the column, increasing the membrane's CO_2 conversion will decrease the required melting area and thus reduce the conversion time.

The main parameters that affect carbonation under microwave-radiated salt mixtures were the CO_2 purity and the impact on capture cycling time.

The parameters that affect the CO_2 purity are the pressure ratio and the column CO_2/HC selectivity. Regarding Fick's law, the ratios of the fluxes for the CO_2 are determined as 60 g carbonate conversion under 10 bar in 11 h cycling time (**Figure 19**).

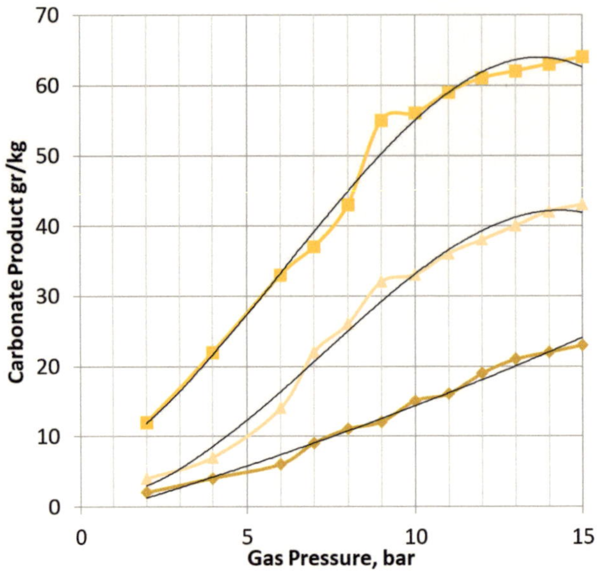

Figure 18.
Effect on CO_2 content with changes in the CO_2/HC of misty flue gas selectivity at pressure ratios of 2–14 bar.

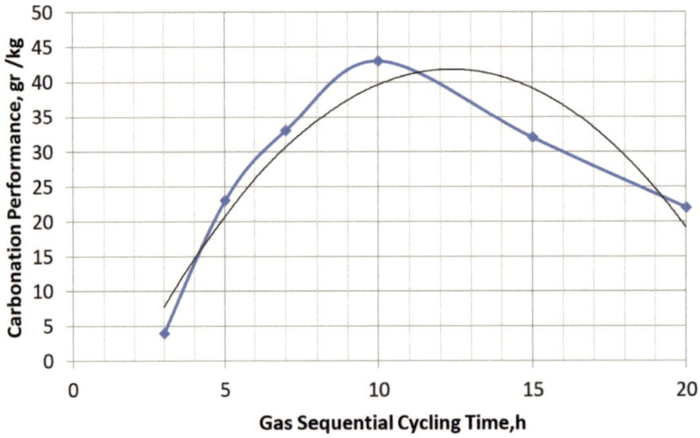

Figure 19.
Change in the capture cost gas sequential cycling time, a function of the three-step column systems under high-pressure carbonation system.

The effect of CO_2 selectivity was not as effective on carbonation due to pressurized carbonation system. By changing to column permeate conditions, the salt/coal char carbonation CO_2, especially for the 10 h column scavenge ring configuration, was significantly improved. However, one disadvantage was that the purity of the CO_2 in the column gas stream was low. It was less than 20% for the rough salt column and less than 80% for scavenger. In the sorption column with salt mixtures, to compete with other CO_2 capture technologies such as chemical absorption and man should also produce high conversion by microwave salt melting and carbonation of CO_2 in coal char.

8. Stabilization quality: geotechnical properties carbonate filler

American Standard (ASTM 3080) experiments were carried out in the fill area. The specimens were tested to determine the geotechnical properties based on the

Component%	Şırnak carbonate	Volcanic slag	Tatvan pumice
SiO_2	33.48	50.50	60.13
Al_2O_3	9.10	14.61	17.22
Fe_2O_3	4.52	24.30	4.59
CaO	22.48	2.30	2.48
MgO	9.80	1.28	2.17
K_2O	2.51	2.51	3.51
Na_2O	1.35	1.35	4.35
Ign. loss.	10.9	0.21	4.12
SO_3	3.32	0.12	0.52

Table 8.
The chemical composition values of Şırnak carbonate, volcanic cinder and Tatvan pumice.

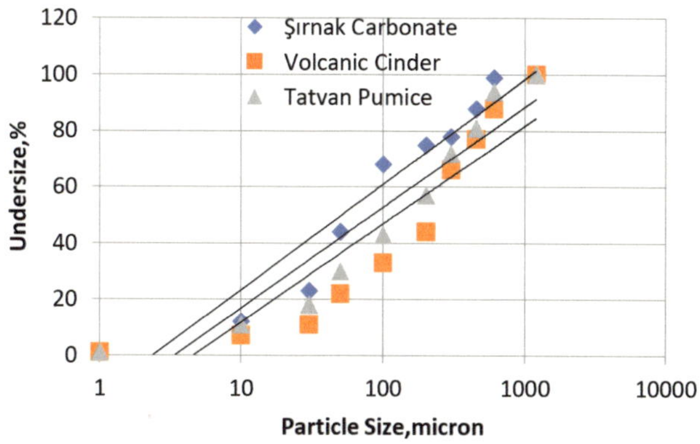

Figure 20.
The grain size distributions of the carbonate output material.

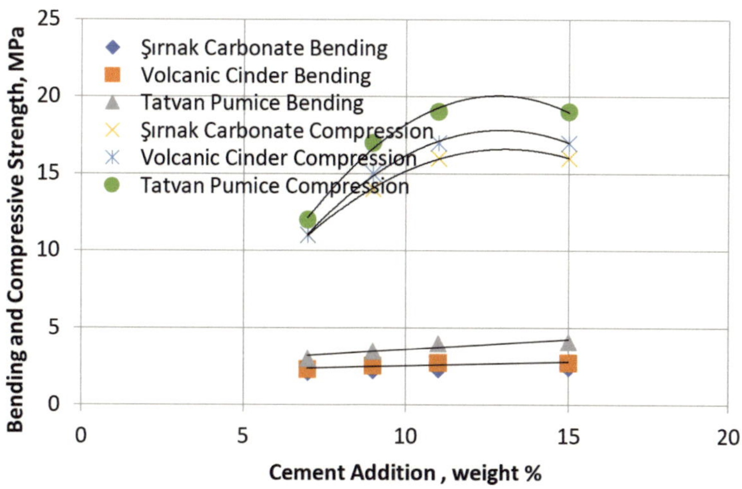

Figure 21.
The tensile and compressive strength of the carbonate cemented block materials.

representative masses in the study area, where the soundings of content are given in **Table 8**. The grain size distributions of the carbonate output material used in stabilization is shown in **Figure 20**. The stabilization cementing blocks of 7 cm cubes were subjected to uniaxial compression strength tests at 28 curing time period. The results for stabilization as bottom layer in the landfill were shown in **Figure 21**. The cemented blocks used Şırnak carbonate and volcanic cinder and pumice were tested as seen from **Figure 21**.

The stability by cement type puzzolane, but local wastes such as fly ash or mid ash of power plants, was used for ground strengthening. The discharge hazardous risk landfill was practiced for contaminated soil area for preventing stabilization and remediation. The strengths of the ground blocks were dispersed to 0.8–1.2 MPa in shear strength and 3.7–4.4 MPa in compression strength. Thus, with the ideal packing, the strength of the mixed cemented blocks produced from these fine fillers and waste mixtures can also reach 11.2 MPa in compression strength in 3.9 MPa in shear strength.

9. Conclusions

This study reveals suitable large-scale operating units in order to achieve the carbonation method as a viable carbonation tool at industrially relevant scales by using fly ash/coal char. Carbonation liquid and gaseous products with fly ash/coal char may change to near 20–45% yield performances with time increase from 1 h to 12 h.

While there is a potential to utilize other types of flue ashes in mineralization, lime or similar alkali can be evaluated to sequester CO_2 allowing clearly significant amounts. There are even researches that succeeded using serpentine and olivine [34–38]. Consequently, the flue gas should be continuously monitored to measure flue gas flow at depleted gas outlet in order to reprocess it.

Other harmful emissions caused by flue gas containing high sulfur (S) and mercury (Hg) content can be eliminated by this method. In that study, results suggested that an appreciable amount of flue gas CO_2 and significant amounts of SO_2 and Hg can be directly captured and mineralized by the fly ash/coal char particles.

Even with progress made so far, to develop an economical method to sequester CO_2 with minerals is still a challenging task, because the process is still relatively slow, and most reactions require high pressure and moderately elevated temperature.

Figure 9 illustrates for the 11 h carbonation period show that cost of carbonation of CO_2 with increasing CO_2 salt carbonation at 65 g by coal char. The results show that the capture cost can be reduced to almost U.S. $20/ton CO_2 avoided when the CO_2 permeability was at high pressure columns 300 bar in the CO_2/N_2 selectivity.

Coal char with CO_2/soot preference of 40–60 mg soot and tar content reduces the carbonation from flue gas to salt reaction below 20%. After the tests, a small quantity of char and soot material was found in the melted salt column scavenger due to coal dissolution reactivity. This material was mainly soot carbon (%99C). Further work is required to determine the soot concentration and compare that with the soot use into gas carbonation.

Author details

Yıldırım İsmail Tosun
Mining Engineering Department, Engineering Faculty, Şırnak University, Şırnak, Turkey

*Address all correspondence to: yildirimismailtosun@gmail.com

References

[1] EEA. European Environmental Agency Waste Statistics. Municipal Waste Management. 2016. Available from: http://www.eea.europa.eu/data-and-maps/indicators/waste-recycling-1/assessment

[2] IEA. World Energy Outlook. Clean Air Report. 2016

[3] Coady AB, Davis JA. CO_2 recovery by gas permeation. Chemical Engineering Progress. 1982;**78**:43

[4] Mazur WH, Chan MC. Membranes for natural gas sweetening and CO_2 enrichment. Chemical Engineering Progress. 1982;**78**:38

[5] Spillman RW, Grace WR. Economics of gas separation membranes. Chemical Engineering Progress. 1989:41

[6] Feron PHM. CO_2 capture: The characterisation of gas separation/removal membrane systems applied to the treatment of flue gases arising from power generation using fossil fuel. IEA Greenhouse Gas R&D Program. Report No. IEA/92/OE8:92-275. Cheltenham, UK; 1992

[7] Hendriks C. Carbon Dioxide Removal from Coal-Fired Power Plants. Dortecht, The Netherlands: Kluwer Academic Publishers; 1994

[8] Van Der Sluis JP, Hendriks CA, Blok K. Feasibility of polymer membranes for carbon dioxide recovery from flue gas. Energy Conversion and Management. 1992;**33**:429

[9] TKI. The Turkish Ministry of Energy. Energy, Dept. Lignite Coal Report; 2015

[10] TTK. The Turkish Ministry of Energy. Energy, Dept. Hard Coal Report; 2011

[11] ASTM 2013, D-3173

[12] ASTM. Standard Test Method for Ash in Biomass E1755-10. PA, USA: ASTM; 2010

[13] ASTM. Standard Test Method for Water Using Volumetric Karl Fischer Titration E203-11. PA, USA: ASTM; 2011

[14] ASTM. Standard Test Method for Ash from Petroleum Products D482-13. PA, USA: ASTM; 2013

[15] Bell DA, Towler BF, Fan M. Coal Gasification and Applications. Oxford: Elsevier Inc.; 2011. ISBN: 978-0-8155-2049-8

[16] Bridgwater AV. Review of fast pyrolysis of biomass and product upgrading. Biomass and Bioenergy. 2012;**38**:68-94

[17] Bridgwater AV. Upgrading fast pyrolysis liquids. Journal Environmental Progress & Sustainable Energy. 2012;**31**:261-268

[18] Bridgwater AV. Fast pyrolysis for bioenergy and biofuels. In: International Congress and Expo on Biofuels & Bioenergy, August 25-27, 2015 Valencia, Spain. 2015. Available from: http://slideplayer.com/slide/9340560/

[19] Çakal GÖ, Yücel H, Gürüz AG. Physical and chemical properties of selected Turkish lignites and their pyrolysis and gasification rates determined by thermogravimetric analysis. Journal of Analytical and Applied Pyrolysis. 2007;**80**(1):262-268

[20] Dahmen N, Henrich E, Dinjus E, Weirich F. The BIOLIQ® bioslurry gasification process for the production of biosynfuels, organic chemicals, and energy. Energy, Sustainability and Society. 2012;**2**(1):3

[21] Donskoi E, McElwain DLS. Approximate modelling of coal pyrolysis. Fuel. 1999;**78**:825-835

[22] Higman C, Burgt M. Gasification. Boston: Elsevier/Gulf Professional Pub; 2003

[23] Jess A, Andresen A-K. Influence of mass transfer on thermogravimetric analysis of combustion and gasification reactivity of coke. Fuel. 2009. DOI: 10.1016/j.fuel 2009.09.002

[24] Kajitani S, Suzuki N, Ashizawa M, et al. CO_2 gasification rate analysis of coal char in entrained flow coal gasifier. Fuel. 2006;**85**:163-169

[25] Kajitani S, Suzuki N, Ashizawa M, Hara S. CO_2 gasification rate analysis of coal char in entrained flow coal gasifier, 1999. Fuel. 2006;**85**(2):163-169

[26] Liu G, Benyon P, Benfell KE, Bryant GW, Tate AG, Boyd RK. The porous structure of bituminous coal chars and its influence on combustion and gasification under chemically-controlled conditions. Fuel. 2002;**79**:617-626

[27] Milne TA, Evans RJ, Abatzoglou N. Biomass gasifier—Tars||: Their nature, formation, and conversion. In: Contract No. NREL/TP-570-25357. Golden, Colorado: National Renewable Energy Laboratory; 1998

[28] Minchener AJ, McMullan JT. Clean Coal Technology. London: IEA Coal Research Ltd; 2007

[29] NBSC, National Bureau of Statistics of China. China Statistical Yearbook. Beijing, China: China Statistics Press; 2014

[30] Öhrman OGW, Weiland F, Pettersson E, Johansson A-C, Hedman H, Pedersen M. Pressurized oxygen blown entrained flow gasification of a biorefinery lignin residue. Fuel Processing Technology. 2013;**115**(2013):130-138

[31] Schora FB. Fuel gasification. In: 152nd Meeting of American Chemical Society, New York. 1967

[32] Schurtz R, Fletcher TH. Pyrolysis and gasification of a sub-bituminous coal at high heating rates. In: 26th Annual International Pittsburgh Coal Conference; 20-23 Sept. 2009

[33] Shadle LJ, Monazam ER, Swanson ML. Coal gasification in a transport reactor. Industrial and Engineering Chemistry Research. 2001;**40**:2782-2792

[34] Sharma A, Saito I, Takanohashi T. Catalytic steam gasification reactivity of hypercoals produced from different rank of coals at 600-775°C. Energy & Fuels. 2008;**22**:3561-3565

[35] Siefert NS. US patent 8,920,526, production of methane-rich syngas from hydrocarbon fuels using multi-functional catalyst/capture agent. 2014. Available from: http://patents.com/US-8920526.html

[36] Tosun YI. Semi-fused salt-caustic mixture leaching of turkish lignites-sorel cement use for desulfurization. In: Proceedings of XIIIth International Mineral Processing Symposium, Bodrum, Turkey. 2012

[37] US EPA. Facts and Figures of Waste Management. 2015. Available from: https://www.epa.gov/sites/production/files/2016-11/panelizersidemswimage2014_0.jpg

[38] Venderbosch RH, Prins W. Entrained flow gasification of bio-oil for syngas. In: Knoef HAM, editor. Handbook Biomass Gasification. 2nd ed. Enschede, The Netherlands: BTG Biomass Technology Group BV; 2012. pp. 219-250

[39] Wei-Biao F, Quing-Hua W. A general relationship between the kinetic parameters for the gasification of coal chars with CO_2 and coal type. Fuel Processing Technology. 2001;**72**:63-77

[40] Eurostat. Energy, transport and environment indicators—2018 edition. In: Environment and Energy Statistical Books. 2018. DOI: 10.2785/94549. Product Code: KS-DK-18-001. Available from: https://ec.europa.eu/eurostat/web/products-statistical-books/-/KS-DK-18-001. ISBN: 978-92-79-96509-8, ISSN: 2363-2372

[41] Weiland F, Nordwaeger M, Olofsson I, Wiinikka H, Nordin A. Entrained flow gasification of torrefied wood residues. Fuel Processing Technology. 2014;**125**:51-58

[42] Wheelock TD. Chemical Cleaning, Coal Preparation. 4th ed. New York: AIME; 1979

[43] Wiktorsson LP, Wanzl W. Kinetic parameters for coal pyrolysis at low and high heating rates—A comparison of data from different laboratory equipment. Fuel. 2000;**79**:701-716

[44] Yoon RH. Advanced Coal Cleaning, Part 2. Coal Preparation. 5th ed. Colorado: AIME; 1991

[45] Chapel DG, Mariz CL, Ernest J. Recovery of CO_2 from flue gases: Commercial trends. In: Proceedings of the Canadian Society of Chemical Engineers, Sakatoon, Saskatchewan, Canada. 1999

[46] Allinson G, Ho MT, Neal PN, Wiley DE. The methodology used for estimating the costs of CCS. In: Proceedings of the Eighth International Conference on Greenhouse Gas Technologies (GHGT-8), Tronheim, Norway. 2006

[47] Shindo Y, Hakuta T, Yoshitome H, Inoue H. Calculation methods for multicomponent gas separation by permeation. Separation Science and Technology. 1985;**20**:445

[48] Burruss R. Geologic sequestration of carbon dioxide in the next 10 to 50 years: An energy resource perspective. In: Prepared for The 10-50 Solution: Technologies and Policies for a Low-Carbon Future, Pew Center on Global Climate Change and the National Commission on Energy Policy, March 25-26, Washington, DC. 2004

[49] Butt DPK, Lackner S, Wendt CH, Park YS, Benjamin A, Harradine DM, et al. A method for permanent disposal of CO_2 in solid form. World Resource Review. 1997;**9**(3):324-336

[50] Davis. Gasification and carbon capture and storage: The path forward. In: Prepared for The 10-50 Solution: Technologies and Policies for a Low-Carbon Future, Pew Center on Global Climate Change and the National Commission on Energy Policy March 25-26, Washington, DC. 2004

[51] Drägulescu C, Tribunescu P, Gogu O. Lösungs gleichgewicht von MgO aus Serpentinen durch Einwirkung von CO_2 und Wasser. Revue Roumaine de Chimie. 1972;**17**(9):1518-1524

[52] DOE. Vision 21, Clean Energy for the 21st Century. U.S. Department of Energy, Office of Fossil Energy DOE/FE-0381. November 1998. Available from: www.fetc.doe.gov/publications/brochures/

[53] Goff F, Guthrie G, Counce D, Kluk E, Bergfeld D, Snow M. Preliminary Investigations on the Carbon Dioxide Sequestering Potential of Ultramafic Rocks. Los Alamos, NM: Los Alamos National Laboratory; 1997; LA-13328-MS

[54] Goldberg PM, Chen Z-Y, O'Connor W, Walters R, Lackner K, Ziock H. CO_2 mineral sequestration studies. In: Paper presented at GlobeEx 2000, August, Las Vegas, NV. 2000

[55] Gunter WD, Perkins EH, McCann TJ. Aquifer disposal of CO_2 rich gases reaction design for added capacity.

Energy Conversion and Management. 1993;**34**:941-948

[56] Hepple R, Benson S. Implications of surface seepage on the effectiveness of geologic storage of carbon dioxide as a climate change mitigation strategy. In: Gale J, Kaya Y, editors. Proceedings of Greenhouse Gas Control Technologies 6 International Conference (GHGT-6), Vol. 1. Elsevier Science Ltd.; 2003. pp. 261-266

[57] IEA. World Energy Outlook; 2012

[58] IPCC. IPCC special report on carbon dioxide capture and storage. In: Metz B, Davidson O, de Coninck HC, Loos M, Meyer LA, editors. Prepared by Working Group III of the Intergovernmental Panel on Climate Change. Cambridge, United Kingdom/New York, NY, USA: Cambridge University Press; 2005. pp. 442

[59] Keeling CD, Whorf TP, Wahlen M, van der Plicht J. Interannual extremes in the rate of rise of atmospheric carbon dioxide since 1980. Nature. 1995;**375**:666-670

[60] Kojima T, Nagamine A, Ueno N, Uemiya S. Absorption and fixation of carbon dioxide by rock weathering. Proceedings of the Third International Conference on Carbon Dioxide Removal, Cambridge Massachusetts, September 9-11. Energy and Conservation Management. 1997, 1996;**38**(Suppl):S461-S466

[61] Lackner KS, Wendt CH, Butt DP, Sharp DH, Joyce EL. Carbon dioxide disposal in carbonate minerals. Energy (Oxford). 1995;**20**(11):1153-1170

[62] Lindeberg E. The quality of a CO_2 repository: What is the sufficient retention time of CO_2 stored underground. In: Gale J, Kaya Y, editors. Proceedings of Greenhouse Gas Control Technologies 6 International Conference (GHGT-6), Vol. 1. Elsevier Science Ltd; 2003. pp. 255-260

[63] Myer L. Sensitivity and cost of monitoring geologic sequestration using geophysics. In: Gale J, Kaya Y, editors. Proceedings of Greenhouse Gas Control Technologies 6th International Conference (GHGT-6), Vol. 1. Elsevier Science Ltd.; 2003. pp. 377-382

[64] O'Connor WK. Investigations into carbon dioxide sequestration by direct mineral carbonation. In: Presentation at Second Meeting of Mineral Sequestration Working Group, November 3, Albany Research Center, Albany, Oregon. 1998

[65] O'Connor WK, Dahlin DC, Nilsen DN, Walters RP, Turner PC. Carbon dioxide sequestration by direct mineral carbonation with carbonic acid. In: Presentation at 25th International Technical Conference on Coal Utilization & Fuel Systems, March 7, Clearwater, Florida; 2000

[66] Oldenburg C et al. Process modeling of CO_2 injection into natural gas reservoirs for carbon sequestration and enhanced gas recovery. Energy & Fuels. 2001;**15**(2):293-298

[67] Pacala S. Global constraints on reservoir leakage. In: Gale J, Kaya Y, editors. Proceedings of Greenhouse Gas Control Technologies 6th International Conference (GHGT-6), Vol. 1. Elsevier Science Ltd.; 2003. pp. 267-272

[68] Plasynski SI, Bose CB, Bergman PD, Dorchak TP, Hyman DM, Loh HP, et al. Carbon mitigation: A holistic approach to the issue. In: Paper presented at the 24th Intl. Tech. Conf. On Coal Utilization and Fuel Systems, March 8-11, Clearwater, FL. 1999

[69] Rubin E, Rao A. Uncertainties in CO_2 capture and sequestration costs. In: Gale J, Kaya Y, editors. Proceedings of Greenhouse Gas Control Technologies 6 International Conference(GHGT-6), Vol. 1. Elsevier Science Ltd.; 2003. pp. 1119-1124

[70] Seifritz W. CO_2 disposal by means of silicates. Nature. 1990;**345**:486

[71] Siegenthaler U, Oeschger H. Biospheric CO_2 emissions during the past 200 years reconstructed by deconvolution of ice core data. Tellus. 1987;**39B**:140-154

[72] Simbeck D. CO_2 capture economics. In: Prepared for The 10-50 Solution: Technologies and Policies for a Low-Carbon Future, Pew Center on Global Climate Change and the National Commission on Energy Policy, March 25-26, Washington, DC. 2004

[73] Solomon S. Carbon Dioxide Storage: Geological Security and Environmental Issues—Case Study on the Sleipner Gas Field in Norway. Oslo, Norway: The Bellona Foundation; 2006. Available from: http://www.bellona.no/artikler/notat_solomon

[74] Stangeland A. A model for the CO_2 capture potential. International Journal of Greenhouse Gas Control. 2007;**1**:2007. Available from: http://www.bellona.no/artikler/notater_stangeland_solomon

[75] Stevens S et al. Sequestration of CO in depleted oil and gas fields: Global 2 capacity, costs, and barriers. In: Williams D, McMullan P, Smith, editors. Proceedings of Greenhouse Gas Control Technologies 5 th International Conference (GHGT-5). 2001. pp. 278-283

[76] Stevens S et al. CO_2 sequestration in deep coal seams: Pilot results and worldwide potential. In: Eliasson R, Wokaun, editors. Proceedings of Greenhouse Gas Control Technologies 4th International Conference (GHGT-4). 1999. pp. 175-180

[77] Torp TA, Gale J. Demonstrating storage of CO_2 in geological reservoirs: The Sleipner and SACS projects. Energy. 2004;**29**:1361-1369

[78] Zweigel P et al. Prediction of migration of CO_2 injected into and underground depository: Reservoir geology and reservoir modeling in the Sleipner case (North Sea). In: Williams D, Mcmullan P, Smith, editors. Proceedings of Greenhouse Gas Control Technologies 5th International Conference (GHGT-5). 2001. pp. 360-365

Chapter 2

Carbon-Based Materials for De-Fluoridation of Water: Current Status and Challenges

Rajeev Kumar and Jyoti Chawla

Abstract

World is facing scarcity of pure and safe drinkable water and third world war would be based on this issue. Recently ground water is excessively used to meet drinking water needs. Water is the principal source of fluoride in daily intake. Excessive fluoride content in ground water due to leaching from fluoride bearing rocks, pose a serious threat worldwide. Concentration of fluoride in drinking water beyond the recommended standards may lead to serious health problems such as skeletal problems, restricted movement, severe anemia and fluorosis. De-fluoridation of water is quite difficult and expensive. Various materials and technologies have been developed to solve this world wide problem. Ion-exchange, precipitation, electro-chemical, reverse osmosis and adsorption are most widely applied methods for de-fluoridation of water. The main highlight of this chapter is to identify and compare carbon-based materials for de-fluoridation of water on the basis of their efficiency, cost and availability. Challenges associated with the development and use of cost effective and environmental friendly materials for de-fluoridation of water have also been discussed.

Keywords: carbon-based materials, de-fluoridation, ground water purification, water contaminates, de-fluoridation techniques, environmental impacts

1. Introduction

Water is an important constituent of our body. It is not only essential for survival but also improves the quality of life. Water is polluted every day by various pollutants or industrial effluents. Fluoride is of the great environmental concern pollutant which contaminates ground water and affects human health. More than 260 million people around the world is affecting by the excess fluoride concentration in groundwater [1]. Only in India, more than 60 million people are at risk of developing fluorosis from fluoride contaminated drinking water [2, 3]. Water is the primary major source of fluoride in daily intake by human beings. The beneficial or detrimental effects of fluoride in water depend on the concentration of fluoride. As per the World Health Organization (WHO), the maximum acceptable limit of fluoride in water is 0.0015 g/L [4]. However, Bureau of India Standards (BIS) has set a limit between 0.0005 and 0.0010 g/L. The acceptable limit varies among countries and lower concentration is recommended for children [5].

Elemental fluoride is more toxic than its oxidized forms and has adverse health effects on human beings as well as on environment. Small concentrations of fluoride in water reduce the incidence of caries, stimulate bone formation and harden the enamel of teeth [6]. When the concentration is beyond the acceptable limit, it causes lesion of the liver, thyroid and endocrine glands, dental and skeletal fluorosis, arthritic symptoms and bone fracture well before the onset of crippling fluorosis, etc. [7–10].

Naturally fluoride originates in ground water due to leaching or dissolution from fluoride bearing rocks. Fluorspar, cryolite, fluorapatite and sellaite are the main fluoride rich rocks that contaminate the ground water after leaching or dissolution [2, 3]. Volcanic ash and rocks are often enriched in fluoride. Fertilizers such as superphosphate, NPK (nitrogen phosphorous potassium) and potash also contain remarkable quantity of fluoride to contaminate ground water [2, 3].

De-fluoridation of water is quite difficult and expensive. Various natural and synthetic materials have been applied to solve this world wide problem. Ion-exchange, precipitation, nano-filtration, electro-chemical, reverse osmosis and adsorption are most widely developed techniques methods for de-fluoridation of water [2, 3, 11]. Among them adsorption is quite effective method because it is easy to operate, needs less space, eco-friendly and cost effective method. Natural, natural modified and synthetic materials have been widely applied as adsorbents for removal of fluoride ions from water [2, 3].

Alumina, aluminum-based materials, manganese dioxide with coated alumina, bauxite, clay and soil, fired clay, carbon-based materials, synthetic resin and biopolymers have been extensively applied as adsorbents for removal of fluoride ions from water [2, 3]. Natural materials /plant waste materials such as rice husk, groundnut shell, saw dust are also most widely used materials used as natural adsorbents.

Carbon-based materials have shown great usefulness for water purification because they exhibit excellent adsorption characteristics and after modifications properties of these materials may be tailored as per requirement. The aim of this chapter is to identify and compare carbon-based materials for de-fluoridation of water on the basis of their efficiency, cost and availability. Challenges associated with the development and use of cost effective and environmental friendly materials and methods for de-fluoridation of water have also been discussed.

2. Background

2.1 Sources of fluoride in water

Groundwater without any physical or chemical treatment is the main source of drinking and other household purposes in most of the countries all over the world. United States, South America, Middle East of Asia, South-East of Africa, India and China are the fluoride affected regions (**Table 1**) [12–23]. However, India and China are the most affected countries worldwide by fluorosis [11]. Fluoride concentration in groundwater depends on the chemical, physical, geological properties of the aquifer. It also depends on the surrounding temperature, intensity of weathering, depth of the wells, porosity and acidity of the soil and rocks, interaction of other chemical elements present in the aquifer [24].

Soil and groundwater are contaminated with fluoride by natural or anthropogenic sources. Naturally soil and groundwater are contaminated by release of fluoride from weathering of the primary minerals fluorite, hydro-geothermal vents, and volcanic activities. However, the use of fluoride bearing fertilizers (fumigants,

Name of country	Sources of fluoride	Amount of fluoride in ground water (mg/L)	References
Brazil	Minerals, fertilizers	0.1–4.7	[16]
Canada	Minerals, fertilizers, rocks	0.1–15.0	[17]
China	Minerals, rocks	2.5–10.3	[18]
Ethiopia	Rocks	0.01–13.0	[19]
Ghana, Ketabasin	Mineral weathering	0–282.2	[20]
India	Geological, and chemical	0.1–16.5	[21]
Kenya	Geological, chemical, rocks	0.1–25.0	[22]
Korea	Geological, and chemical	0–48.8	[23]
Mexico	Geological, and chemical	0.5–7.5	[24]
Pakistan	Minerals, fertilizers, rocks	0.11–22.8	[25]
Sri Lanka	Minerals, fertilizers, rocks	0.1–4.3	[26]
USA, Wisconsin	Geological, and chemical	0.01–7.6	[2]

Table 1.
Fluoride concentration in groundwater in different country.

pesticides, aluminum phosphate fertilizer) and burning of coal increases the fluoride level in soil and groundwater in terms of anthropogenic ways.

The main primary source of fluoride is fluoride rich minerals or rocks. Ground water is contaminated from fluoride due to leaching from fluoride bearing minerals, cryolite (Na_3AlF_6), fluorspar (CaF_2), villiaumite (NaF), sellaite (MgF_2), topaz ($Al_2(SiO_4)F_2$), fluorapatite ($Ca_5(PO_4)_3F$) and fluoride bearing rocks, conglomerate, schist, killas, silexite, granite, sandstone, and gneiss [25]. Dissolution of fluorite mineral from the host rocks increases in alkaline medium. pH 7.6–8.6 is the most favorable range of dissolution of fluorite mineral [2]. Fluoride shows a constructive connection with sodium and bicarbonate while it shows inverse connection with calcium in water.

2.2 Health effects of fluoride

Exposure to fluoride more than the recommended value is associated with a number of health issues. Dental fluorosis, skeletal fluorosis, neurological problems, thyroid problems and other health problems are the main health issues associated with high exposure of fluoride [5, 8, 11].

Hypo-mineralization of tooth enamel due to excessive fluoride intake is known as dental fluorosis. Fluoride decreases the availability of free calcium ions in the mineralization environment. It is characterized by discoloration of the teeth or formation of pits in the teeth depending upon dose, age, and time of exposure. Faint white lines, white mottled patches, brown discoloration, brittle, pitted and rough enamel are the various stages of initial to severe fluorosis (**Figure 1**). The effects of dental fluorosis may not appear in case of already developed teeth [11, 18, 25]. It does not mean that fluoride intake in adults is within the safety limit. Only in India more than 60 million people are suffering from fluorosis. More than 85% children had dental fluorosis.

Bone and joint deformations is known as skeletal fluorosis due to excessive intake of fluoride (**Table 2**). The bones may become hardened, thicken and less elastic causes severe pain, impaired joint mobility and increasing the risk of fractures. Sporadic pain, stomach-ache, muscle weakness and stiffness of joints are the early symptoms of skeletal fluorosis. Hardening and calcifying of the bones (osteosclerosis) are the next stage and spine, muscles, nervous system and major joints damage are the final stage of

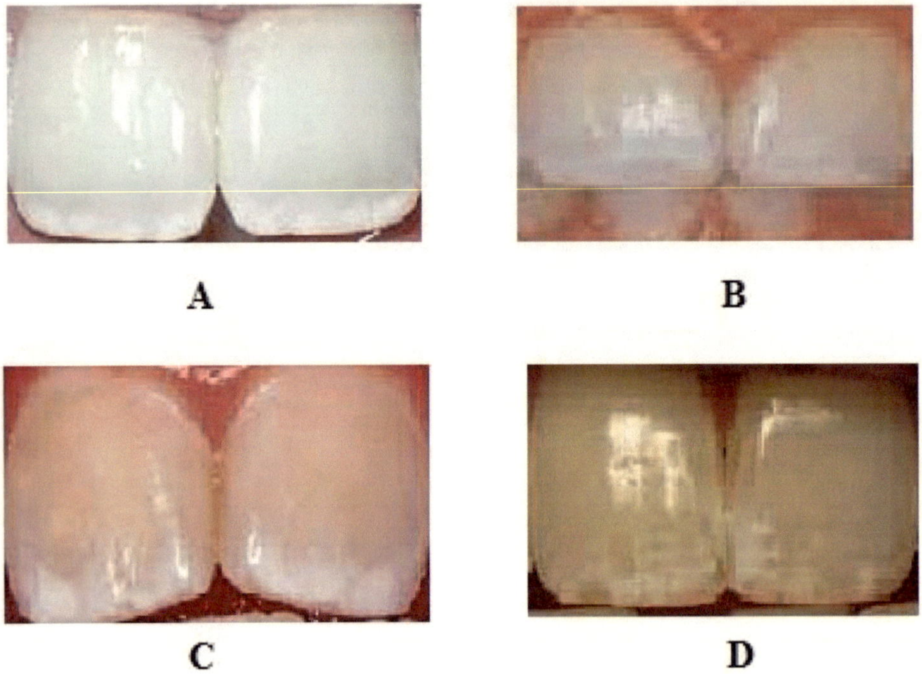

Figure 1.
Dental fluorosis: (A) normal, (B) mild, (C) moderate, (D) severe conditions.

S. No.	Name of standard agency	Permissible limit (mg/L)	Health effects
1	World Health Organization (WHO)	0.6–1.5	<0.5 = dental caries
2	Bureau of Indian Standards (BIS)	0.6–1.5	0.5–1.5 = optimum dental health
3	US Public Health Standards	0.8	1.5–4.0 = dental fluorosis
4	Indian Council of Medical Research (ICMR)	1.0	<4.0 = dental skeletal and crippling fluorosis

Table 2.
Permissible limit of fluoride in drinking water and associated effects on teeth.

skeletal fluorosis. More than 35% children had skeletal fluorosis. In both of the cases no treatment exists, thus prevention is better than cure. Long time exposure of fluoride also changes the DNA structure and may causes cancer [7].

Long time exposure of fluoride to pregnant women may affect cognitive ability for the child in the future. Excessive exposure to fluoride was associated with low intelligent quotient. Recently, fluoride was found as a neurotoxin affects the child development. Excessive exposure of fluoride may also causes high blood pressure, cardiovascular problems, cardiac insufficiency, heart failure, acne, skin problems, reproductive issues, thyroid dysfunction [7–10].

2.3 De-fluoridation techniques

Exposure of fluoride has various adverse effects. Therefore its monitoring and removal is very important. A number of techniques are available for the removal of fluoride from water. Physical, chemical and biological methods are applied for defluoridation of drinking water. Ion exchange, precipitation, membrane process,

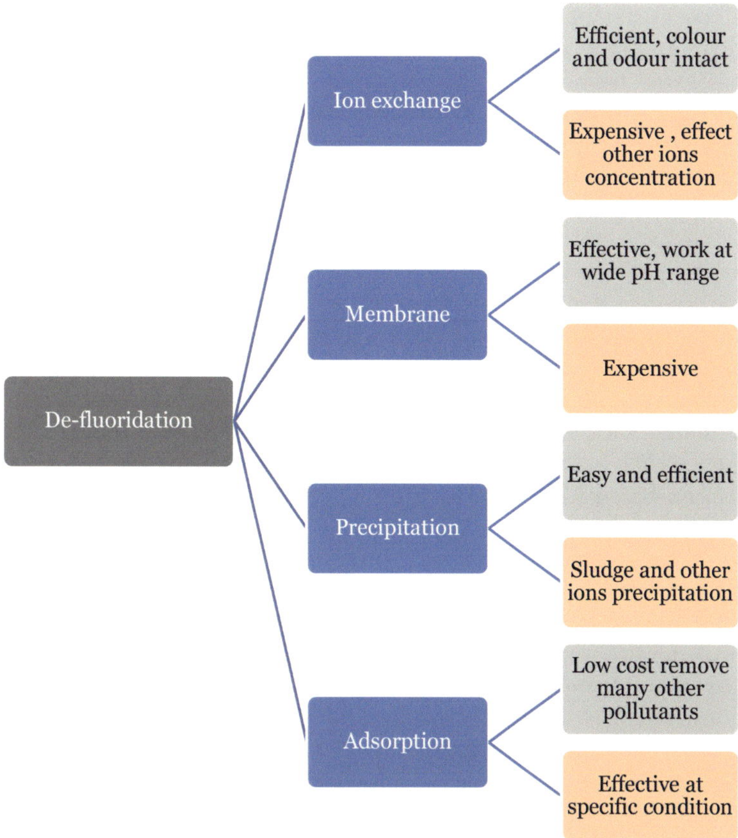

Figure 2.
Schematic diagram for some important defluoridation techniques.

adsorption and phyto- and bioremediation are most promising methods have been applied for defluoridation of drinking water (**Figure 2**) [2, 3, 11, 12].

In ion exchange method synthetic anionic and cationic exchanger resin are applied for defluoridation of water. Lewatit MIH59, Tulsion A-27, Ambalite TRA400, Deaceodite FFIP, Polyanion (NCL), anionic exchange resin and polystyrene resin, sulfonated saw-dust carbon, Wasoresin IR, cationic exchanger resins are most widely used ion exchangers for removal of fluoride from water.

In precipitation method fluoride in water gets precipitated out in the form of fluorapatite. The insoluble fluorapatite gets separated out from the aqueous phase. Nalgonda technique, contact precipitation and MgO, $Ca(OH)_2$ and $NaHSO_4$ mixtures are most widely used methods for precipitation of the fluoride from water.

Membrane process involves the compound specific permeability membrane with specific pore size. Reverse osmosis, electro-dialysis and nano-filtration are the method used for defluoridation of water using membranes [26].

Adsorption is one of the most promising methods widely applied for defluoridation of the water. It is cheap, needs less space and easy to handle method. Natural and synthetic materials are used as adsorbent for the removal of fluoride from water [26].

3. Discussion

Different adsorbent materials are reported in literature for effective removal of fluoride from water. Carbon-based adsorbents are often used for the removal of contaminants from water as their surface can be tailored as per the nature of contaminant.

3.1 Carbon-based materials as adsorbents for de-fluoridation

Carbon-based materials have unique properties such as high stability, large surface area, and high strength [2]. These materials can combine with other elements or materials to form strong covalent bond. Due to unique properties and versatile nature they have been widely used in different fields. Researchers have used different types of carbon-based materials for removal of fluorine from water. Various carbonaceous materials can be used to prepare activated carbon via chemical or physical activation. Carbon-based nanomaterials are also fast emerging materials for defluoridation of the water because of their very small size, high surface area, remarkable electrical conductivity, unique structural dimensions, high mechanical strength, and high efficiency [27]. In this chapter all the reported studies are reviewed and applicability of carbon-based adsorbents in different conditions are analyzed.

3.2 Natural and modified activated carbon/carbon-based waste

Abe et al. investigated the defluoridation of water by using various carbon-based charcoal in order to get the best adsorption capacity. The maximum adsorption capacity was found to be in following order bone char > coal charcoal > wood charcoal > carbon black > petroleum coke [28]. Defluoridation of water using activated carbon depends on various factors, such as pH, solubility, polarity, pore size distribution, molecular size of the adsorbate, surface area, surface functional groups, and other ions in solution. Microporous activated carbons have high surface areas shows high adsorption capacity for the adsorption of low molecular weight compounds [29]. Up to 85% of fluoride ions from water was removed by powdered activated charcoal at pH 2 adsorbent dose of 2.4 g/100 mL, stirring rate of 60 rpm and contact time of 120 min, which was very effective than the untreated charcoal [30]. Considering low pH vlaues, it is due to availability of the large number of H^+ ions. Up to 17, 10, and 82% defluoridation of water were achieved by using commercial activated carbon, carbon black, and bone char, respectively at 30°C at a fixed contact time of 3 h. Sreenivasulu et al. investigated defluoridation of water by using activated carbon prepared from Umbles of Prangos Pabularia Lindl [31]. In batch studies adsorption was completed within 90 min for 0.0028–0.0076 g/L fluoride concentration. Removal percentage increases up to 85% with increase in the concentration up to 7 g/1 at pH of 8.75 and temperature of 57°C. Singh et al. investigated defluoridation of water by using activated charcoal of wheat husk (AC) and alum treated fly ash (ATF) obtained from thermal power station in Agra city. About 68.6% defluoridation of water was achieved at 2 g AC/100 ml of ground water at pH 6.8 and temperature 25°C in 24 h [32]. Charcoal is produced by slow heating of wood, coal, bone, lignite, nutshells, and petroleum residues, low cost waste materials such as coconut shell, fruit waste, rice-husk, saw dust, tree bark, and cotton waste in the absence of oxygen. Arulanantham et al. were investigated defluoridation of water by using coconut shell carbon (CSC) and commercial activated carbon (CAC). Defluoridation of water from dilute aqueous solutions was very effective when coconut shell carbon impregnated with alum. Adsorption capacity of wet carbon was three times more than that the dry carbon. Coconut shell carbon has certain specific advantages over alumina for defluoridation of water at pH 5–8 [33]. Various animal bone char and its activated forms are widely used as adsorbent for defluoridation of water. Removal percentage depends on the initial fluoride concentration of solution, adsorbent dose, pH of the solution and contact time [34]. High defluoridation of water at pH 7.0 was observed by normal cow bone char and under different conditions such as CO_2 environment, NO_2 environment, Al doped cow bone char [35]. Maximum adsorption capacity was found to be 7.32–31 mg/g. Highest adsorption capacity was found to be for Al doped cow bone char (31 mg/g). Al doped cow bone char was prepared at 700°C temperature [36]. Adsorption capacity of bone chars are altered by the functional groups on the surface. Zuniga-Muro et al. were investigated defluoridation of water by doped two different cerium precursors (Ce^{3+} and Ce^{4+}) onto cattle bone char. The results showed that a significant enhancement in the adsorption capacity of fluoride on to the Ce^{4+} modified bone char composite from 5.47 to 13.6 mg/g at pH 7 [37]. Zhu et al. were investigated defluoridation of water by cattle bone char after modification with different aluminum salts ($AlCl_3$, $AlNO_3$, $NaAlO_2$, $Al_2(SO_4)_3$) [38]. About 97% removal was achieved onto $AlCl_3$ modified bone char at pH 7. Moreover, maximum adsorption capacity was found to be 6.8 mg/g by using bone char modification using $AlCl_3.6H_2O$ [39]. However, bone char pre-treatment with $Ca(OH)_2$, $FeCl_3$, $CaCl_2$ and $MgCl_2$ were less effective for defluoridation of water with maximum adsorption capacity 4.4, 1.56, 5.1 and 4.2 mg/g, respectively. Maximum adsorption capacity for unmodified bone char was found to be 7.32 mg/g. Rojas-Mayorga et al. were investigated defluoridation of water by doped aluminum sulfate onto bovine bone char (pyrolyzed at 700°C) on packed bed micro-columns [40]. Maximum adsorption capacity for bovine bone char (pyrolyzed at 700°C) at pH 7 was found to be 3.3–18.5 mg/g. The removal efficiency has significantly improved by doping bone char with different metals (Fe_xF_y, $Al(OH)_xF_y$, and CaF_2). The maximum adsorption capacity for aluminum sulfate doped bone char at pH 7 was found to be 31 mg/g. Chatterjee et al. were investigated defluoridation of water by aluminum sulfate with calcium oxide to chemically treat carbonized bone meal (a mixture of chicken and cattle bones). The maximum adsorption capacity increases for chemically treat carbonized bone meal from 14 to 150 mg/g [41].

Ma et al. were investigated defluoridation of water by using granular activated carbon coated with manganese oxides. Removal effieciency was three times higher than uncoated granular activated carbon at pH 3 [42]. Rao et al. were used activated carbon of bergera koenigh carbon (BKC), batavia orange carbon (BOC) and *Raphanus sativus* carbon (RSC) for defluoridation of carbon. 0.0040 g/L fluoride content was reduced to permissible level by using theses adsorbents at pH 6, 1 g/L adsorbent dose and at 30 min of contact time [43]. Roy and Das were investigated defluoridation of water by using activated carbon prepared from tea waste (ACTW). About 99.59% removal was achieved at adsorbent dose 1.0 g temperature of 60°C and contact time of 70 min [44].

Ramos et al. investigated defluoridation of water by using plain and alumina-impregnated activated carbons. Alumina-impregnated activated carbons were prepared by stirring with an aluminum nitrate solution at a fixed pH. Calcinations of the alumina-impregnated activated carbons were done under nitrogen at temperatures 300°C. Adsorption capacity of the adsorbent depends on the pH of the impregnating solution and the temperature of calcinations. Alumina-impregnated activated carbons showed 3–5 time's high adsorption capacity than the plain activated carbon [45]. Janardhana et al. investigated defluoridation of water by using zirconium impregnated activated charcoals. It also showed 3–5 time's high adsorption capacity than the plain activated carbon [46]. Dahiya and Kaur investigated defluoridation of water by using coconut coir pith carbon (CPC). The activated form of CPC was obtained by carbonization in presence of sulfuric acid without any chemical treatment as well as after impregnation with different alum dose. 78.8% defluoridation of water from standard fluoride solution of 0.0025 g/L after contact period of 12 h was achieved on 10 g/L CPC impregnated with 2% alum. Removal capacity of adsorbent decreased with increase in initial fluoride concentration and with decrease in dose of adsorbent [47].

Gupta et al. studied the defluoridation of water by using waste carbon slurries from fuel-oil energy generators. Solid was activated by heating in air at 450°C. Activated material was washed with sodium hydroxide solution and fluoride free water to remove ash and other contaminates. The final material which contains 92.0% carbon, 0.45% aluminum and 0.6% iron was dried at 100°C. Defluoridation of water and regeneration of material were pH-dependent with optimum pH 7.6 [48]. Thermally activated biosorbents prepared from banana (*Musa paradisiaca*) peel and coffee (*Coffea arabica*) husk were used as adsorbents for defluoridation of water [49]. Maximum adsorption capacity was achieved at pH of 2, 24 g/250 mL at 13 h contact time for banana peel and 18 g/250 mL at 3 h contact time for coffee husk. The real water samples were collected in consultation with WRDA (water resource development authority) office at Hawassa city, Ethiopia. The concentrations of fluoride in flour factory, poultry, and Lake Hawassa sites were found to be 0.0012, 0.0011 and 0.0067 g/L, respectively. The prepared adsorbents were used for the same water. The removal percentage was from 80 to 84% [49]. Karuga et al. investigated defluoridation of water by using activated fish swim bladder-derived porous carbon (FBPC). Maximum adsorption capacity was achieved at pH of 6, adsorbent dose of 5.0 g/L and contact time of 50 min. It follows pseudo second-order kinetic and Langmuir isotherm models for adsorption process [50].

3.3 Carbon-based pristine and functionalized nanomaterials

Carbon-based nanomaterials have also attracted considerable attention in the recent years for defluoridation of water with higher uptake capacity. Li et al. were investigated defluoridation of water by using graphene a single flat two-dimensional (2D) atomic sheet of carbon. Maximum adsorption capacity (17.65 mg/g) was obtained at 0.0025 g/L initial fluoride concentration and 25°C temperature [51]. Dongre was investigated defluoridation of water by using fabricated chitosan doped graphite novel composite (FCDGNC). Langmuir maximum adsorption capacity was found to be 37.9 mg/g at pH 6.5. It follows pseudo second order model. FCDGNC was regenerated and reused better upon five cycles [52]. Roy et al. were investigated defluoridation of water by using reduced graphene oxide. Chemical reduced graphene oxide (CRGO) can be prepared from 400 mg of graphene oxide dispersed in 400 mL deionized water by means of 30 min ultra sonication with ammonium hydroxide and hydrazine hydrate. Biochemical synthesis of graphene oxide can be done by tea polyphenol. Tea polyphenol reduced graphene oxide (TPGO) can be prepared by 50 mg graphene oxide powder added in the tea solution and sonicated for 30 min at 363 K in a nitrogen atmosphere. The removal percentage was 94.22% by TPGO whereas it was 87.4% in case of CRGO. Langmuir adsorption isotherm was the best fitted model for both of the adsorbents. The kinetic follows pseudo second order model. TPGO can be regenerated by using 1% sodium hydroxide solution and reused for defluoridation of water [53]. Aligned carbon nanotubes (ACNT) were prepared by decomposition of xylene, catalyzed by ferrocene [54]. ACNT adsorbs 4.5 mg/g fluoride from 0.0015 g/L fluoride at pH 7. Adsorption capacity increases with increasing the acidity or positive charge on the surface of ACNT. Adsorption capacity under the similar conditions for carbon nanotubes, typical soil, g-Al_2O_3 and activated carbon increase in following manner activated carbon < g-Al_2O_3 < soil < CNT. Haghighat et al. were investigated defluoridation of water using single and multi-wall carbon nanotubes (SWCNs and MWCNs) [55]. The study showed that 58% removal efficiency in 70 min was obtained at pH 5, 0.0010 g/L fluoride initial concentration by using 0.5 g/L single-wall carbon nanotube. However, 54% removal efficiency in 70 min was obtained at pH 5, 0.0010 g/L fluoride initial concentration by using 0.5 g/L multi-wall carbon nanotube. In acidic condition adsorption capacity was increased. SWCNs showed high removal efficiency in comparison to two types of fine powder and 150 mesh activated alumina in optimum conditions.

Tang et al. investigated the defluoridation of water by using novel hydroxy-apatite decorated with carbon nanotube composite (CNT-HAP) [56]. Maximum adsorption capacity using CNT-HAP composite for removal of fluoride was found to be 11.05 mg/g at pH 6. Freundlich model showed the best fitted isotherm model. Regression coefficient showed that it follows pseudo-second-order kinetic model. Defluoridation of water was also investigated by using alumina-impregnated carbon nanotubes [57, 58]. Carbon nanotubes were synthesized by the pyrolysis of a propylene hydrogen mixture with Ni particles as the catalyst. The product was ball-milled and then stirred with aluminum nitrate solution at 500°C under nitrogen for 2 h. A sponge-like alumina supported on carbon nanotubes was obtained which was ground and sieved to appropriate particle size. The adsorption capacity of the alumina-impregnated carbon nanotubes was found to be very high (13.5 times) compared to AIC-300 (4 times) g-Al_2O_3 and IRA-410 polymeric resin. Li et al. were investigated defluoridation of water by using carbon nanotubes supported on alumina. Maximum adsorption capacity 9.6 mg/g was achieved at pH of 6 at alumina loading of 30 wt%. The adsorption follows Freundlich isotherm model. It follows second order rate equation [59]. Gupta et al. were investigated defluoridation of water by using a micronanohierarchal web (MiNaHiWe) consisting of activated carbon fibers (ACF) and carbon nanofibers (CNF), impregnated with Al. Aluminum carbon nanofibers (CNF) was applied for treating the wastewater at pH 5–8 [60].

3.4 Comparison of carbon-based materials for defluoridation of water in pH range 6.0–7.5

Carbon-based materials such as aligned carbon nanotubes (ACNT), single and multi-wall carbon nanotubes (SWCNs and MWCNs), hydroxyapatite decorated with carbon nanotube composite (CNT-HAP), alumina-impregnated carbon nanotubes (Al-CNT), charcoal, activated charcoal of various materials, bone char of various animals, graphene, chemical and bio-reduced i.e. chemical reduced graphene oxide (CRGO), tea polyphenol reduced graphene oxide (TPGO) and fabricated chitosan doped graphite novel composite (FCDGNC) are used for defluoridation of water. Maximum adsorption capacity of these materials depends on the pH of the solution, adsorbent dose, contact time, initial fluoride concentration and temperature. It is very difficult to identify the best adsorbent among the following for defluoridation of the water because the removal conditions are different (**Table 3**). However, the studies reveal that chemically treated carbonized bone meal is the best adsorbent for defluoridation of water. The maximum adsorption capacity of chemically treated (CT-CBM) is very high compared to other used adsorbents under the slightly different conditions. The maximum adsorption capacity of the used adsorbents at pH 6.0–7.5 follows the following order (**Figure 3**):

CT-CBM (150) > FCDGNC (37.9) > Al-S-BC (31.0) > TPGO (28.72) > CRGO (18.22) > CBM (14.0) > CNT-HAP (11.05) > Al-CNT (9.6) > CBC (7.32) > ACNT (4.5) > activated charcoal of materials (ACM) (~2.0) > charcoal of materials (CM) (~1.0).

Comparison of maximum adsorption capacity of different materials was also studied by various groups under the similar conditions. Adsorption capacity of

Carbon-based materials	Max adsorption capacity (mg/g)	pH	Isotherm/kinetics	Reference
ACNT	4.5	7.0	Langmuir/ pseudo-second-order	[54]
CNT-HAP	11.05	6.0	Langmuir/ pseudo-second-order	[56]
Al-CNT	9.6	6.0	Langmuir/ pseudo-second-order	[59]
FCDGNC	37.9	6.5	Langmuir/ pseudo-second-order	[52]
CRGO	18.22	7.1	Langmuir/ pseudo-second-order	[53]
TPGO	28.72	7.1	Langmuir/ pseudo-second-order	[53]
Activated banana peel	0.39	6.3	Langmuir/ pseudo-second-order	[49]
Activated coffee husk	0.41	6.3	Langmuir/ pseudo-second-order	[49]
Bagasse AC	1.15	6.0	Langmuir/ pseudo-second-order	[61]
Saw dust AC	1.73	6.0	Langmuir/ pseudo-second-order	[61]
Wheat straw AC	1.93	6.0	Langmuir/ pseudo-second-order	[61]
Fish swim bladder carbon	1.43	6.0	Langmuir/ pseudo-second-order	[50]
Cow bone char (CBC)	7.32	7.0	Langmuir/ pseudo-second-order	[36]
Cow bone char under CO_2 environment	5.92	7.0	Langmuir/ pseudo-second-order	[36]
Al sulfate doped (Al-S-BC)	31.0	7.0	Langmuir/ pseudo-second-order	[37]
Carbonized bone meal (CBM)	14.0	6.1	Langmuir/ pseudo-second-order	[41]
Chemically treated (CT-CBM)	150.0	6.1	Langmuir/ pseudo-second-order	[48]

Table 3.
Maximum adsorption capacity of carbon-based materials for defluoridation of water.

carbon nanotubes, typical soil, g-Al_2O_3 and activated carbon follows following order CNT > soil > g-Al_2O_3 > activated carbon under the similar conditions.

However, SWCNs showed high removal efficiency of fluoride in comparison to MWCNs which was also very high to two types of fine powder and 150 mesh activated alumina in optimum conditions.

The adsorption capacity of the alumina-impregnated carbon nanotubes was found to be very high (13.5 times) compared to AIC-300 (4 times) g-Al_2O_3 and IRA-410 polymeric resin under similar conditions.

In similar manners the maximum adsorption capacity of charcoals was found to be in following order bone char > coal charcoal > wood charcoal > carbon black > petroleum coke under optimum conditions.

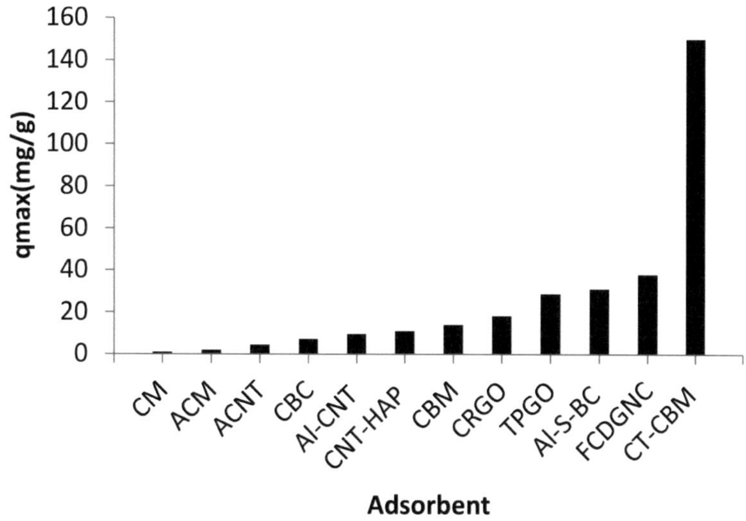

Figure 3.
Comparison of maximum adsorption capacity of carbon-based materials for defluoridation of water.

4. Conclusions

More than 260 million people around the world are exposed to excess fluoride concentration in groundwater which is the prime reason of dental and skeletal fluorosis. There is need to identify the feasible cost effective and eco-friendly materials for defluoridation of water at community level. Various natural and synthetic materials have been explored for defluoridation of water. The adsorption capacity was found to be very poor for pristine natural alternatives. However, they are cost effective, easily disposable and environmental friendly. Chemically treated carbonized bone meal is found to be best adsorbent for defluoridation of water. The maximum adsorption capacity of chemically treated (CT-CBM) was found very high when compared with other used adsorbents under the slightly different conditions. Carbon-based nanomaterials like graphene and its fabricated forms are also most widely used adsorbents for the defluoridation of water. Cost and disposal of graphene and nanomaterials is a very crucial issue while proposing these materials for defluoridation of water. Future studies should be focused to explore more low cost and eco-friendly materials that promise good adsorption capacity for fluorine and can be used to treat drinking water in normal conditions.

Acknowledgements

The authors are grateful to the administration and management of Manav Rachna International Institute of Research and Studies, Faridabad, Haryana, India, for providing infrastructure and other support for preparation of this chapter and other ongoing research.

Author details

Rajeev Kumar* and Jyoti Chawla
Manav Rachna International Institute of Research and Studies, Faridabad, Haryana, India

*Address all correspondence to: rajeevkumar.fet@mriu.edu.in

References

[1] Amini M, Mueller K, Abbaspour KC, Rosenberg T, Afyuni M, Møller KN, et al. Statistical modeling of global geogenic fluoride contamination in ground waters. Environmental Science & Technology. 2008;**42**:3662-3668

[2] Mohapatra M, Anand S, Mishra BK, Giles DE, Singh P. Review of fluoride removal from drinking water. Journal of Environmental Management. 2009;**91**:67-77

[3] Wambu EW, Ambusso WO, Onindo C, Muthakia GK. Review of fluoride removal from water by adsorption using soil adsorbents—An evaluation of the status. Journal of Water Reuse and Desalination. 2015;**6**(1):1-29

[4] WHO (World Health Organization). Guidelines for Drinking-Water Quality: Incorporating First Addendum to Third Edition. Geneva: World Health Organization; 2006. 375 p

[5] Erdal S, Buchanan SN. A quantitative look at fluorosis, fluoride exposure, and intake in children using a health risk assessment approach. Environmental Health Perspectives. 2005;**113**:111-117

[6] Fung K, Zhang Z, Wong J, Wong M. Fluoride contents in tea and soil from tea plantations and the release of fluoride into tea liquor during infusion. Environmental Pollution. 1999;**104**:197-205

[7] Kanduti D, Sterbenk P, Artnik B. Fluoride: A review of use and effects on health. Materia Socio Medica. 2016;**28**:133-137

[8] Czarnowski W, Krechniak J, Urbańska B, Stolarska K, Taraszewska-Czarnowska M, Muraszko-Klaude A. The impact of waterborne fluoride on bone density. Fluoride. 1999;**32**(2):91-95

[9] Dey S, Giri B. Fluoride fact on human health and health problems: A review. Medical & Clinical Reviews. 2016;**1**:1-6

[10] Ullah R, Zafar MS, Shahani N. Potential fluoride toxicity from oral medicaments: A review. Iranian Journal of Basic Medical Sciences. 2017;**20**:841-848

[11] Ayoob S, Gupta AK. Fluoride in drinking water: A review on the status and stress effects. Critical Reviews in Environmental Science and Technology. 2006;**36**:433-487

[12] Mirlean N, Roisenberg A. Fluoride distribution in the environment along the gradient of a phosphate-fertiliser production emission (southern Brazil). Environmental Geochemistry and Health. 2007;**29**(3):179-187

[13] Desbarats AJ. On elevated fluoride and boron concentrations in groundwaters associated with the Lake Saint-Martin impact structure, Manitoba. Applied Geochemistry. 2009;**24**:915-927

[14] Genxu W, Guodong C. Fluoride distribution in water and the governing factors of environment in arid North-West China. Journal of Arid Environments. 2001;**49**:601-614

[15] Tekle-HaimanotR,MelakuZ,KloosH, Reimann C, Fantaye W, Zerihun L, et al. The geographic distribution of fluoride in surface and groundwater in Ethiopia with an emphasis on the Rift Valley. Science of the Total Environment. 2006;**367**:182-190

[16] Yidana SM, Yakubo BB, Akabzaa TM. Analysis of groundwater quality using multivariate and spatial analyses in the Keta basin, Ghana. Journal of African Earth Sciences. 2010;**58**:220-234

[17] Datta PS, Deb DL, Tyagi SK. Stable isotope (180) investigations on the processes controlling fluoride contamination of groundwater. Journal of Contaminant Hydrology. 1996;**24**:85-96

[18] Gaciri SJ, Davies TC. The occurrence and geochemistry of fluoride in some natural waters of Kenya. Journal of Hydrology. 1993;**143**:395-412

[19] Chae GT, Yun ST, Mayer B, Kim KH, Kim SY, Kwon JS, et al. Fluorine geochemistry in bedrock groundwater of South Korea. Science of the Total Environment. 2007;**385**:272-283

[20] Valenzuela-Vásquez L, Ramírez-Hernández J, Reyes-López J, Sol-Uribe A, Lázaro-Mancilla O. The origin of fluoride in groundwater supply to Hermosillo City, Sonora, Mexico. Environmental Geology. 2006;**51**:17-27

[21] Farooqi A, Masuda H, Kusakabe M, Naseem M, Firdous N. Distribution of highly arsenic and fluoride contaminated groundwater from East Punjab, Pakistan, and the controlling role of anthropogenic pollutants in the natural hydrological cycle. Geochemical Journal. 2007;**41**:213-234

[22] Young SM, Pitawala A, Ishiga H. Factors controlling fluoride contents of groundwater in north-central and northwestern Sri Lanka. Environmental Earth Sciences. 2011;**63**(6):1333-1342

[23] Ozsvath DL. Fluoride concentrations in a crystalline bedrock aquifer, Marathon County, Wisconsin Marathon county, Wisconsin. Environmental Geology. 2006;**50**(1):132-138

[24] Feenstra L, Vasak L, Griffioen J. Fluoride in groundwater: Overview and evaluation of removal methods. In: International Groundwater Resources Assessment Centre Report nr. SP 2007-1. 2007. pp. 1-21

[25] He J, An Y, Zhang F. Geochemical characteristics and fluoride distribution in the groundwater of the Zhangye basin in the north western China. Journal of Geochemical Exploration. 2013;**135**:22-30

[26] Waghmare SS, Arfin T. Fluoride removal from water by various techniques: Review. International Journal of Innovative Science, Engineering & Technology. 2015;**2**:560-571

[27] Premathilaka RW, Liyanagedera ND. Fluoride in drinking water and nanotechnological approaches for eliminating excess fluoride. Journal of Nanotechnology. 2019:1-15

[28] Abe I, Iwasaki S, Tokimoto T, Kawasaki N, Nakamura T, Tanada S. Adsorption of fluoride ions onto carbonaceous materials. Journal of Colloid Interface Science. 2004;**275**: 35-43

[29] Sivarajasekar N, Paramasivan T, Muthusaravanan S, Muthukumaran P, Sivamani S. Defluoridation of water using adsorbents—A concise review. Journal of Environment and Biotechnology Research. 2017;**6**(1): 186-198

[30] Tembhurkar AR, Dongre S. Studies on fluoride removal using adsorption process. Journal of Environmental Science and Engineering. 2006;**48**:151-156

[31] Sreenivasulu A, Sumdaram EV, Reddy MK. Defluoridation studies using activated carbon prepared from umbles of *Parangos pavularia* Lindl. Journal of Industrial Pollution Control. 1999;**15**(2):24-27

[32] Singh RP, Singh Y, Swaroop D. Defluoridation of groundwater in Agra

City using low cost adsorbents. Bulletin of Environmental Contamination and Toxicology. 2000;**65**(1):120-125

[33] Arulanantham A, Ramcrishna TV, Balsubramaian N. Studies on fluoride removal by coconut shell carbon. Journal of Environmental Protection. 1992;**12**:531-532

[34] Alkurdi SSA, Al-Juboori RA, Bundschuh J, Hamawand I. Bone char as a green sorbent for removing health threatening fluoride from drinking water. Environment International. 2019;**127**:704-719

[35] Rojas-MayorgaCK,Silvestre-AlberoJ, Aguayo-VillarrealIA,Mendoza-CastilloDI, Bonilla-Petriciolet A. A new synthesis route for bone chars using CO_2 atmosphere and their application as fluoride adsorbents. Microporous and Mesoporous Materials. 2015;**209**:38-44

[36] Rojas-Mayorga CK, Bonilla-Petriciolet A, Silvestre-Albero J, Aguayo-Villarreal IA, Mendoza-Castillo DI. Physico-chemical characterization of metal-doped bone chars and their adsorption behavior for water defluoridation. Applied Surface Science. 2015;**355**:748-760

[37] Zuniga-MuroN,Bonilla-PetricioletA, Mendoza-Castillo D, Reynel-Ávila H, Tapia-Picazo J. Fluoride adsorption properties of cerium-containing bone char. Journal of Fluorine Chemistry. 2017;**197**:63-73

[38] Zhu H, Wang H, Wang G, Zhang K. Removal of fluorine from water by the aluminum-modified bone char. In: 2010 International Conference on Biology, Environment and Chemistry, IPCBEE; Singapore. Vol. 1. 2011. pp. 455-457

[39] Nigri EM, Cechinel MAP, Mayer DA, Mazur LP, Loureiro JM, Rocha SD, et al. Cow bones char as a green sorbent for fluorides removal from aqueous solutions: Batch and fixed-bed studies. Environmental Science and Pollution Research. 2017;**24**(3):2364-2380

[40] Rojas-Mayorga CK, Bonilla-Petriciolet A, Sánchez-Ruiz FJ, Moreno-Pérez J, Reynel Ávila HE, Aguayo-Villarreal I, et al. Breakthrough curve modeling of liquid-phase adsorption of fluoride ions on aluminum-doped bone char using micro-columns: Effectiveness of data fitting approaches. Journal of Molecular Liquids. 2015;**208**:114-121

[41] Chatterjee S, Mukherjee M, De S. Defluoridation using novel chemically treated carbonized bone meal: Batch and dynamic performance with scale-up studies. Environmental Science and Pollution Research. 2018;**25**:18161-18178

[42] Ma Y, Wang SG, Fan M, Gong WX, Gao BY. Characteristics and defluoridation performance of granular activated carbons coated with manganese oxides. Journal of Hazardous Materials. 2009;**168**:1140-1146

[43] Rao VS, Chakrapani C, Babu CS, Rao KS, Rao MN, Sinha D. Studies on sorption of fluoride by prepared activated Kaza's carbons. Der Pharma Chemica. 2011;**3**:73-83

[44] Roy S, Das P. Assessment on the defuoridation using novel activated carbon synthesized from tea waste: Batch, statistical optimization and mathematical modeling. Journal of Industrial Pollution Control. 2016:1-24

[45] Leyva-RamosR,Ovalle-TurrubiartesJ, Sanchez-Castillo MA. Adsorption of fluoride from aqueous solution on aluminum-impregnated carbon. Carbon. 1999;**37**:609-617

[46] Janardhana C, Rao G, Shatish R, Kumar P, Kumar V, Madhav V. Study on defluoridation of drinking water using zirconium ion impregnated activated

charcoals. Indian Journal of Chemical Technology. 2007;**14**(4):350-354

[47] Dahiya S, Kaur A. Studies on removal of fluoride by coconut coir pith carbon. Indian Journal of Environmental Protection. 1999;**19**(11):68-71

[48] Gupta VK, Ali I, Saini VK. Defluoridation of wastewaters using waste carbon slurry. Water Research. 2007;**41**(15):3307-3316

[49] Getachew T, Hussen A, Rao VM. Defluoridation of water by activated carbon prepared from banana (*Musa paradisiaca*) peel and coffee (*Coffea arabica*) husk. International Journal of Environmental Science & Technology. 2015;**12**:1857-1866

[50] Karuga J, Jande YAC, Kim HT, King'ondu CK. Fish swim bladder-derived porous carbon for defluoridation at potable water pH. Advances in Chemical Engineering and Science. 2016;**6**:500-514

[51] Li Y, Zhang P, Du Q, Peng X, Liu T, Wang Z, et al. Adsorption of fluoride from aqueous solution by grapheme. Journal of Colloid and Interface Science. 2011;**363**:348-354

[52] Dongre RS. Biosorption of fluoride from water by fabricated chitosan doped graphite novel composite. Research & Development in Material Science. 2018;7(1):1-9

[53] Roy S, Manna S, Sengupta S, Ganguli A, Goswami S, Das P. Comparative assessment on defluoridation of waste water using chemical and bio-reduced graphene oxide: Batch, thermodynamic, kinetics and optimization using response surface methodology and artificial neural network process. Safety and Environmental Protection. 2017;**1**(1):221-231

[54] Li Y-H, Wang S, Zhang X, Wei J, Xu C, Luan Z, et al. Adsorption of fluoride from water by aligned carbon nanotubes. Materials Research Bulletin. 2003;**38**:469-476

[55] Haghighat GA, Dehghani MH, Nasseri S, Mahvi AH, Rastkari N. Comparison of carbon nonotubes and activated alumina efficiencies in fluoride removal from drinking water. Indian Journal of Science and Technology. 2012;**5**(3)

[56] Tang Q, Duan T, Li P, Zhang P, Wu D. Enhanced defluoridation capacity from aqueous media via hydroxyapatite decorated with carbon nanotube. Frontiers in Chemistry. 2018;**6**. Article 104

[57] Li Y, Wang S, Cao A, Zhao D, Zhang XC, Luan Z, et al. Adsorption of fluoride from water by amorphous alumina supported on carbon nanotubes. Chemical Physics Letters. 2001;**350**:412-416

[58] Li YH, Wang S, Zhang X, Wei J, Xu C, Luan Z, et al. Removal of fluoride from water by carbon nanotube supported alumina. Environmental Technology. 2003;**24**:391-398

[59] Li Y, Wang S, Zhao X, Wei J, Xu C, Luan Z, et al. Removal of fluoride from water by carbon nanotube supported alumina. Environmental Technology. 2008;**24**:391-398

[60] Gupta AK, Deva D, Sharma A, Verma N. Adsorptive removal of fluoride by micro-nanohierarchal web of activated carbon fibers. Industrial & Engineering Chemistry Research. 2009;**48**:9697-9707

[61] Yadav AK, Abbassi R, Gupta A, Dadashzadeh M. Removal of fluoride from aqueous solution and groundwater by wheat straw, sawdust and activated bagasse carbon of sugarcane. Ecological Engineering. 2013;**52**:211-218

Chapter 3

Graphene- and Graphene Oxide-Bounded Metal Nanocomposite for Remediation of Organic Pollutants

Brajesh Kumar

Abstract

Nanotechnology is one of the most interesting areas concerned with consumer products including cosmetics, household appliances, electronics, textiles, and food production as well as in medical products. Environmentally benign, economical, practical, and efficient processes for the synthesis of graphene (rGO)-/graphene oxide (GO)-bounded metal nanoparticles and their use for the remediation of organic pollutants (dyes, pharmaceutical wastes, pesticides, etc.) have been increasingly important goals in the chemical community from economic, safety, and environmental points of view. In this chapter, various strategies have successfully demonstrated the synthesis of graphene-/graphene oxide-bounded metal nanoparticles using various natural sources (plant extracts, biomolecules, polysaccharide, alcohols, etc.) and their applications in environmental remediation.

Keywords: green synthesis, graphene, graphene oxide, organic dye, XRD, adsorption, photocatalytic effect, ecofriendly

1. Introduction

In 1959 American theoretical physicist Richard P. Feynman popularization through his lecture "There is a plenty of room at the bottom," revolutionize a new era technology, where manipulation of the properties of a device is possible at atomic, molecular and macromolecular scales called nanotechnology [1]. Nanoparticles, having particle sizes ranging between 1 and 100 nanometers, have a very surprising history, their synthesis is not limited to a modern laboratory, they have existed in nature for a long time, and their use can be traced to ancient times; for example, the application of metal nanoparticles as color pigments in luster and glass appeared in Mesopotamia during the ninth century. Nanosized metal particles have unique chemical and physical properties that are significantly different from those of the bulk equivalents [2]. Now, the deliberate manipulation of the matter at the aforementioned scale allows the creation of materials with unique (and unusual) optical, chemical, photoelectrochemical, electronic, and biological properties that can be applied in catalysis, biosensing, imaging, medicine, and other important fields [3, 4].

Eco-friendly methods for nanoparticle synthesis using plants or plant extracts, fruits, agricultural wastes, microorganisms, enzymes, fungi, starch, sucrose, and maltose have been suggested as possible alternatives to chemical and physical methods, and biological approaches to the synthesis of nanoparticles are attracting research attention. Biological molecules can undergo highly controlled, hierarchical assembly, which makes them suitable for the development of reliable and eco-friendly processes for metal nanoparticle synthesis. Furthermore, these methods are cheap and rapid and efficiently produce single atoms or molecules with a wide variety of shapes (spheres, prisms, or plates) in the nanoscale level [4, 5]. Phytosynthesis of metallic nanostructures has become a more suitable alternative technique than physical and chemical methods because they are simple, nontoxic, rapid, cost-effective, eco-friendly, and sustainable. In the other hand, it has been widely recognized that carbon-based nanomaterials such as carbon nanotubes, carbon nanofibers, active carbon, and graphene can be used in a lot of important applications like sensing, catalysis, conversion, energy storage, polymer composites, and pollution prevention; all these properties are strongly coupled to carbon's structural conformation and, thus, its hybridization state [6].

In the past few years, carbonaceous and carbon-based nanomaterials have been attracting a significant research interest because of their exclusive characteristics, including good biocompatibility, nontoxicity, high mechanical/thermal properties, and relative ease of functionalization [7]. Among other nanomaterials, graphene has become a rising star in condensed matter physics and material research due to its exclusive mechanical, physical, and chemical features. The term "graphene" was proposed in 1986 to define single atomic layers of graphite. Since its discovery as important allotropes of carbon, graphene has become one of the most advanced carbon nanomaterials. It is a two-dimensional single sheet sp^2 hybridized of carbon atoms arranged in a hexagonal close-packed honeycomb like a crystal lattice through σ- and π-bonds, and it also has high specific surface area (2,630 $m^2 g^{-1}$) [8]. Specifically, the chemical oxidation-exfoliation-reduction of graphene to obtain graphene oxide (GO), which has abundant oxygen-containing functional groups (epoxy, carbonyl, carboxyl, hydroxyl, etc.), allows the creation of the perfect surface to anchor organic nanoparticles for enhancement of wider applications (**Figure 1**) [9]. Because of its novel and unique properties, tremendous fundamental and technological studies have been stimulated. Graphene has a unique band structure in which the conduction band and valence band just touch each other, forming an exactly zero bandgap semiconductor (also known as a semimetal) [10]. Since the first successful production in 2004 by Andre Geim and his coworkers at

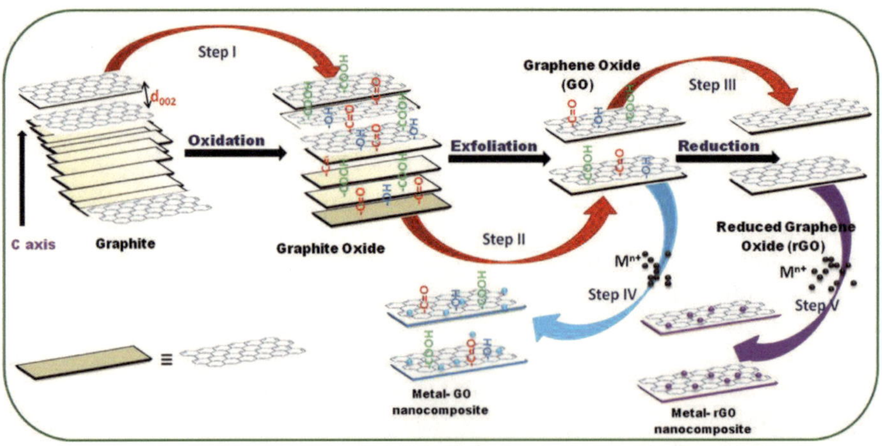

Figure 1.
Schematic representation for the synthesis of graphene oxide, graphene (rGO), and its metal nanocomposite.

Manchester University, graphene have become a "rising star" material thanks to its unique properties such as large theoretical specific surface area, high thermal conductivity, good biocompatibility, optical transmittance, good electrical conductivity, high elasticity, chemical and electrochemical inertness, and easy surface modification; for these reasons graphene can be used in field effect transistors, sensors, transparent conductive films, clean energy storage, conversion devices, catalysis, photocatalysis, nanocomposites, solar cells, drug delivery, etc. [11].

In the last few decades, a vast number of synthetic organic compounds, including textile dyes, pharmaceutical compounds, and pesticides, have been found in wastewater and surface/groundwater due to wastewater discharged from various industries and domestic households. Most of the pharmaceutical compounds and dyes are highly polar, of low volatility, and extremely resistant to biodegrading processes; these constitute a potential risk for the aquatic and terrestrial organism and also a collateral generation of microbial resistance. For this reason, the need of the improvement of the decontamination efficiency of organic pollutants using eco-friendly techniques is increasing with time [12, 13]. Ozonization, photo-fenton, biodegrading processes, and photocatalysis have been widely used techniques for organic compound degradation; all these methods show a high rate of effectiveness [14]. Semiconductor and noble metal nanoclusters in nanometer size display a lot of interesting optical, electronic, chemical, and biological size−/shape-dependent properties; in terms of pollutant degradation, the noble metal nanoparticles exhibit increased photochemical activity because of their unusual electronic properties and high surface [15].

Over the last years, nanotechnologies have been implemented in almost all branches of human activities, including wastewater treatment. Numerous nanomaterials are utilized to eliminate inorganic and organic species from wastewater effluents, in many cases even more efficiently than the conventional adsorbents. These include nanoparticles of metals, metal oxides, carbon nanotubes, graphite, fullerenes, carbon-based nanocomposites, etc. [16]. On the other hand, graphene oxide enhances the possibilities for the construction of graphene-involved catalyst and graphene-based hybrid nanocomposites with inorganic species, such as noble metal nanoparticles in order to photodegrade pollutants (**Figure 1**). The photodegrade activity is promoted by the high specific surface area (2,630 $m^2 g^{-1}$), enhanced active sites, large delocalized π-electron systems, good chemical stability, unique electronic properties of graphene-based materials, and excellent transparency; this means it is an excellent adsorbent for organic and inorganic pollutants [17]. For example, graphene/graphene oxide could be used in combination with common photocatalysts (TiO_2, ZnO), leading to the enhancement of absorptivity of pollutants. This means a new generation of photocatalyst systems used as an adsorbent for wastewater handling purposes [18].

2. Graphene oxide and graphene synthesis

The trend to produce graphene and graphene nanoplatelets in bulk quantities centers on chemical exfoliation of graphite, which is more advantageous than other methods. Ideally, the graphene material should have just one layer, but recently graphene with two or more layers are being investigated with the same interest. One of the most developed techniques to obtain graphene with high yields single layer powder following a modified Hammers method, which consists of the initial oxidation of graphite to water dispersible graphite oxide, which results in an increase in the interlayer spacing between the oxygen-containing graphene layers. The oxygen-containing functional groups make GO sheets highly hydrophilic and render them prone to swelling quickly and therefore easy to be dispersed in water [19].

This process must be followed by mechanical or chemical exfoliation of graphite oxide to graphene oxide sheets and the final reduction to graphene (**Figure 1**) [4, 20]. Also, the reduction of graphite oxide, using thermal or chemical procedures, to obtain graphene, appropriately named reduced graphene oxide, has been a challenging research area. For a long time, hazardous, toxic, and explosive reducing agents have been used, as in the synthesis of metallic nanoparticles [21]. In order to avoid the use of these chemicals, some authors recommend the use of biomolecules such as ascorbic acid, reducing sugars, casein, and plant extracts; generally they contain a lot of polyphenols which are converted to their corresponding oxidized quinone forms with the power to reduce graphene oxide [22].

3. Graphene oxide and graphene-metal nanocomposite synthesis

Graphene and graphene oxide have a plate-like structure with a large specific area and fantastic properties, making the sense to use these kinds of materials as a substrate for the deposition of inorganic nanoparticles like silver, gold, copper, palladium, iron, cobalt, zinc, tin, etc. Several techniques which are used to decorate graphene and graphene oxide are employed, for example, a solution-based method, electrodeposition, thermal evaporation, photochemical, and solventless bulk synthesis, among others; in any case, the type of solvent, nature and concentration of metal precursor, the presence of a dispersing and/or reducing agent, time, and temperature are factors that must be controlled to obtain a narrow particle size distribution [23, 24]. The hydrophilicity and exfoliation of graphene oxide have offered exciting opportunities to build nanocomposites and to form strong physical interactions with small molecules and polymers [24].

Nanoparticles have been produced by chemical strategies for a long time; in this process harsh, toxic, and expensive chemicals (e.g., hydrazine, sodium borohydride, dimethylformamide, and ethylene glycol) could be used as non-environmental friendly reducing agents; for this reason, the goal of the green chemistry is to foster the use of eco-friendly solvent medium and reducing and stabilizing agents [25]. Current research in phytosynthesis using plant extracts have shown that they are a good alternative for fast and nontoxic production of nanoparticles; this is thanks to the antioxidant activity of several plants which have a high content of polyphenols (e.g., flavonoids, flavones), carotenoids, reducing sugars, terpenoids, glutathiones, and metallothioneins, among other molecules, that could act as reducing and capping agents in the nanoparticle formation and replace toxic chemicals [26]. The phytosynthesis of metallic nanoparticles involves three main steps: (i) selection of the solvent medium, (ii) selection of environmentally benign reducing agent, and (iii) selection of nontoxic substances for nanoparticle stability [27].

Nanocatalysis is a recent field of research in nanoscience; most of the studies involve the use of spherical or undefined-shape nanoparticles with a small size, as an alternative to conventional materials [28]. The combination of graphene or graphene oxide with metallic nanoparticles has opened a new field in environmental remediation through the increase of photocatalytic degradation activity against organic pollutants due the ability to absorb light and enhance heterogeneous catalysis [4, 29]. The organic pollutants discharged by industries can remain in the environment because they have high solubility in water and high stability to light, temperature, and biological degradation. Recently, biogenic synthesis of graphene- or graphene oxide-bounded metallic nanocomposite for the photodegradation of organic compounds in wastewater, such as dyes, pharmaceutical and personal care products (PPCPs), and pesticides (**Figure 2**), has received much attention due to the toxicity thereof to human beings and the environment.

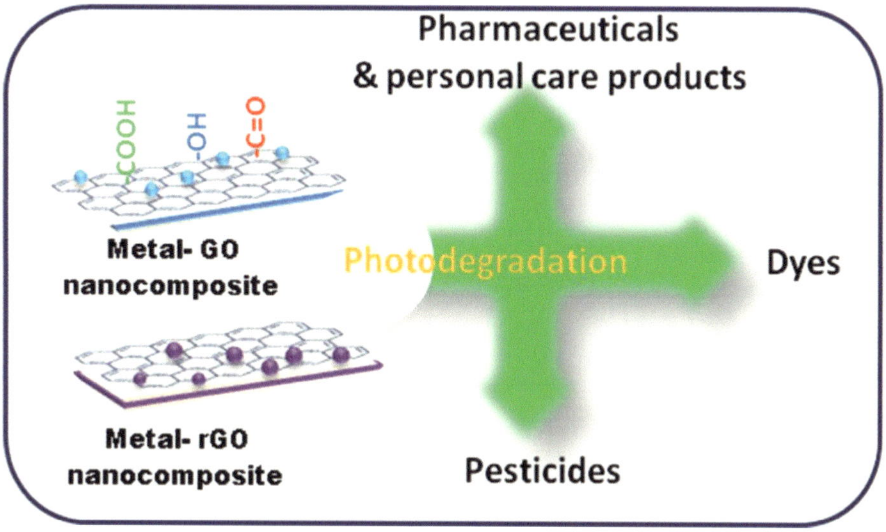

Figure 2.
Utilization of graphene oxide/graphene-metal nanocomposite for the photodegradation of various categories of organic pollutants.

4. Photocatalytic degradation mechanism for organic pollutants

Recently, in order to degrade organic dye compounds, many researchers around the world have studied on the photocatalyst material which is usually a wide bandgap semiconductor in the presence of solar energy. For practical applications, development of efficient, reproducible, and cost-effective visible light-induced photocatalyst is desirable for the large-scale production of catalyst. Recently, many research papers are dedicated to the plasmonic photocatalysis by using noble metals as silver [30], carbonaceous-based (nanotubes, fullerenes, graphene, etc.) [31], and semiconductor-carbonaceous composite based photocatalysts [32] and their structure (to include size, shape, crystallinity, porosity, contact form, etc.). The principals behind the heterogeneous photocatalysis are based on a complex sequence of reactions which create electron–hole pairs under photo-illumination. Briefly photocatalyst absorbs UV and/or visible light irradiation from sunlight, which is abundantly available as natural irradiation energy or an illuminated light source as shown in **Figure 3**. The electrons (e^-) in the valence band (VB) of the photocatalyst are excited to the conduction band (CB), while the holes are left in the valence band. This, therefore, creates the missing negative charges/negative (e^-) and positive electron–hole (h^+) pairs. This stage is referred to the semiconductor's "photoexcited" state, and the energy difference between the valence band and the conduction band is known as the "bandgap." This must correspond to the wavelength of the light for it to be effectively absorbed by the photocatalyst. The electron–hole pairs that are generated in this way migrate toward the surface where they can initiate redox reactions with adsorbates. Photogenerated reactions with the positive holes are linked to oxidation of hydroxyl groups or water molecules to produce hydroxyl (•OH) radicals and reactions with electrons to dissolved oxygen to reduce superoxide radical anion ($•O_2^-$), which then reacts with H^+ to form $•HO_2$ and H_2O_2, followed by rapid decomposition to •OH. Degradation/disinfection of toxic pollutants operates through the formation of •OH, which have very low reaction selectivity. These •OH radicals drive the oxidation of pollutants to complete mineralization, i.e., less reactive pollutants [28, 33].

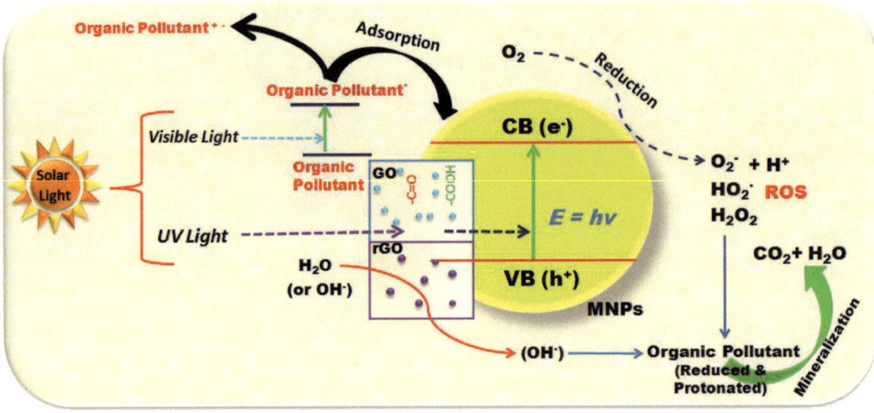

Figure 3.
Mechanism for the photocatalytic degradation of organic pollutants.

5. Photocatalytic degradation of dyes

Over the recent decades, graphene- or graphene oxide-bounded metallic nanocomposite-based photocatalysts have attracted great attention in water purification because of their high photosensitivity, environmental friendliness, stability, nontoxic nature, and low cost. Carbonaceous material is well known as an effective adsorbent due to hydrophobic interaction, π-π interactions, hydrogen-bonding interactions, and electrostatic and dispersion interactions, and it has large specific surface area and also acts as electron scavengers due to their large electron storage capacity. **Figure 4** shows the structural formula for the salts of different toxic organic dyes. Vizuete et al. reported on the formation of a silver-graphene (Ag-G) nanocomposite through the reduction of Ag^+ and graphene oxide using berry extract of Mortiño (*Vaccinium floribundum* Kunth). The as-prepared Ag-G nanocomposite showed an enhanced photocatalytic activity for the degradation of methylene blue (k = 0.0163283 min^{-1}) and methylene orange (k = 0.0140985 min^{-1}) in an aqueous medium under sunlight irradiation [34]. Chandu et al. reported in

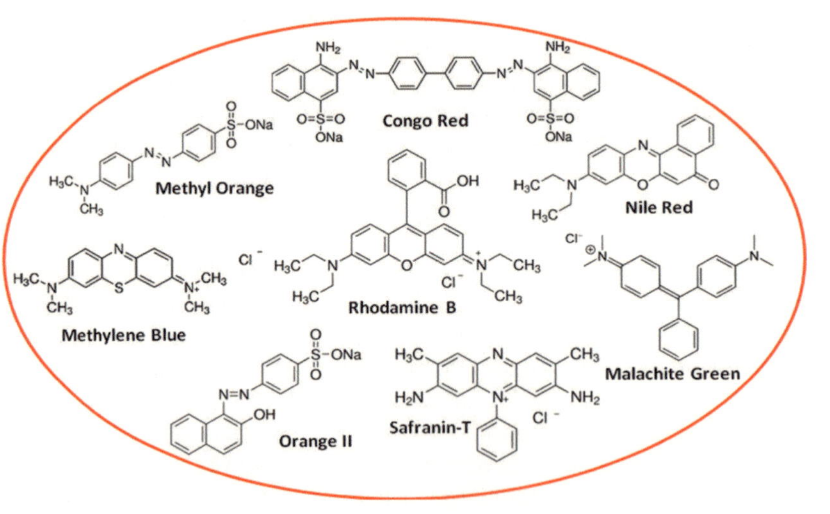

Figure 4.
The structural formula of salt for different toxic organic dyes.

situ synthesis of a nanocomposite of silver decorated on reduced graphene oxide sheets using betel leaf extract as a stabilizing and reducing agent. The sheet structure of reduced graphene oxide and uniformly distributed 28 nm silver nanoparticles exhibited good photocatalytic efficiency (95% in 2 h) against methylene blue in sunlight [35]. In another paper, Chandu et al. efficiently synthesized graphene oxide sheets decorated with silver nanoparticles (CRG-Ag nanocomposite) in 12 h using custard apple leaf extract, and as-synthesized CRG-Ag nanocomposite showed excellent photocatalytic efficiency of 96% in 2 h under sunlight using methylene blue (5 mg/L) as a model pollutant [36]. Maryami et al. used *Abutilon hirtum* leaf extract as a reducing and capping agent for the synthesis of silver supported on the reduced graphene oxide (Ag/RGO). The crystalline nature of the Ag/RGO nanocomposite was identified using XRD analysis and employed as a catalyst for the reduction of organic dyes such as 4-nitrophenol (4-NP), Congo red (CR), and rhodamine B in the shortest possible time in water at mild conditions [37].

Chandra et al. explained that the photodegradation of eosin, methylene blue, and rhodamine B is almost 80% in the presence of graphene-Mn_2O_3 nanocomposite [38]. In other hand, Benjwal et al. elaborated the enhanced photocatalytic degradation of methylene blue by reduced graphene oxide-metal oxide (TiO_2/Fe_3O_4)-based nanocomposites with efficiency (~100% within 5 min) [39]. Nenavathu et al. has synthesized functionalized graphene oxide (FGS)/ZnO nanocomposites with enhanced photocatalytic activity toward the photodegradation of the safranin T dye in aqueous solution via an economical, facile, and simple solution route, followed by calcination under environmental conditions [40]. Xiong et al. investigated the effect of modification of reduced graphene oxide with crystalline copper species. The copper species acted as an electronic relay, passing the excited electrons from the reduced graphene oxide to the adsorbed oxygen. The continuously generated reactive oxygen species led to the degradation of rhodamine B under visible light irradiation [41]. Fu and co-workers prepared a magnetically separable $ZnFe_2O_4$/graphene nanocomposite photocatalyst by a facile one-step hydrothermal method. The photocatalyst exhibited 88% of methylene blue degradation by adding H_2O_2 at 5 min and reached up to 99% at 90 min after irradiation of visible light. It serves a dual function as a photoelectrochemical degrader and a generator of hydroxyl radicals via photoelectrochemical decomposition of H_2O_2 [42].

Lui et al. synthesized hierarchical TiO_2 nanoflowers using a solvothermal reaction. The graphene-wrapped hierarchical TiO_2 nanoflowers showed improved performance of degradation of methylene blue under UV irradiation. This enhanced photodegradation is attributed to the excellent ability of transport of the charge, that is, electron as graphene has excellent conductivity which in turn suppresses the recombination of the electron-hole pair and provides more preferential adsorption sites for methylene blue which in turn enhances the degradation of methylene blue [43]. Lv et al. reported a quick and facile microwave-assisted synthesis method to produce ZnO-reduced graphene oxide hybrid composites and studied the photocatalytic performance in the degradation of methylene blue. The results showed that reduced graphene oxide plays an important role in the enhancement of photocatalytic performance. It was shown that the ZnO-rGO composite with 1.1% wt rGO achieved a maximum degradation efficiency of 88% in a neutral solution under UV light as compared to pure ZnO (68%) due to increased light absorption and reduced charge recombination [44]. Thangavel et al. first reported the synthesis and photocatalytic behavior of β-tin tungstate-reduced graphene oxide (β-$SnWO_4$-rGO) nanocomposites. The degradation efficiency of 55 and 60% were achieved by β-$SnWO_4$ alone, whereas in the presence of rGO, the photocatalytic degradation efficiency was found to be increased up to 90 and 91% in methylene orange and rhodamine B, respectively [45].

Ternary nanocomposites comprising Ag-Cu_2O supported by glucose-reduced graphene oxide with enhanced stability and visible light photocatalytic activity were synthesized by Sharma et al. via a facile and green approach using Benedict's solution and glucose solution at room temperature. The resulting Ag-Cu_2O/rGO nanocomposites showed excellent photocatalytic efficiency for the photodegradation of methyl orange, and the degradation rate was higher than the pristine Cu_2O and Cu_2O/rGO NCs. [46]. Xiong et al. reported an excellent visible light photocatalytic performance for graphene modified with gold nanoparticles in degrading non-biodegradable dyes like rhodamine B, methylene blue, and orange II in the water. The rate constants were calculated to be about 8.7×10^{-3} min^{-1}, 4.1×10^{-2} min^{-1}, and 9.4×10^{-4} min^{-1} for rhodamine B, methylene blue, and orange II, respectively. It confirmed that graphene modified with gold photocatalyst has four important features in comparison to the TiO_2: (1) high adsorption ability toward organic dyes, (2) strong p-p interaction with dye chromophores, (3) efficient photosensitized electron injection, and (4) slow electron recombination. These features make the graphene modified with gold composite a good candidate for photo-assisted degradation of dye pollutants in water [47].

Palladium-decorated ZnS/rGO nanocomposites prepared by a coprecipitation method showed significant visible light-induced photocatalytic activity toward the degradation of indigo carmine dye. The highest photocatalytic activity was 1.0% Pd-ZnS/rGO sample for degradation of 20 ppm indigo carmine ($k = 2.19 \times 10^{-2}$ min^{-1}) [48]. Kurt et al. prepared platinum and palladium-decorated reduced graphene nanosheet having antioxidant activity; IC_{50} values are 46.1 and 90.2 μg/mL, respectively, based on the ABTS method and 80.2 and 143.7 μg/mL according to the DPPH method. It also showed photocatalytic elimination of fuchsine and indigo carmine dyes under light irradiation [49]. Isari et al. prepared a ternary nanocomposite of Fe-doped TiO_2 decorated on reduced graphene oxide using simple sol-gel method. The DRS results of the photocatalysts showed a narrowing bandgap by the introduction of Fe ions to the titania framework. The maximum removal of rhodamine B is 91% after 120 min under solar illumination by using 0.6 g Fe-doped TiO_2/rGO nanocomposite containing 3% Fe and 5% rGO, with an initial pH of 6 [50]. El-Shafai et al. fabricated two nanocomposites, namely, graphene oxide-iron oxide (GO-Fe_3O_4) and graphene oxide-iron oxide-zirconium oxide (GO-Fe_3O_4@ZrO_2), and characterized their identity by using different spectroscopic techniques. It was observed that newly fabricated GO-Fe_3O_4 (3.66 eV) and GO-Fe_3O_4@ZrO_2 (3.20 eV) nanocomposites have the advantage of smaller bandgaps than GO (4.00 eV), which result in increased adsorption capacity and photocatalytic effects. Results confirmed that RhB is efficiently adsorbed over the surface of graphene oxide (~93%) and has a significant inhibitory effect against *E. coli* bacteria [51].

The hydrothermal synthesis and photocatalytic activity of SnS-reduced graphene oxide nanocomposites for the photodegradation of malachite green in water were being investigated by Wang et al.. The experimental results indicated that the as-synthesized nanocomposites showed excellent sunlight-excited photocatalytic activity of malachite green in water with quasi-kinetic rate constant in the range of ~0.32–0.69 h^{-1} and it increased to ~0.60–1.82 h^{-1} by using H_2O_2 synthetically [52]. Ye et al. prepared CdS-graphene and CdS-carbon nanotube nanocomposites by a hydrothermal method and studied them as photocatalysts for the evolution of hydrogen and the degradation of methyl orange under visible light irradiation. The incorporation of graphene or carbon nanotube into CdS significantly enhanced the photocatalytic activities for both reactions. It confirmed that the rate of photodegradation of methylene orange was the highest for the CdS-graphene and it is 7.9 times higher than that of CdS alone and 1.8 times higher than that of the CdS-carbon nanotube composite [53]. Recently, Ansari et al. for the first time fabricated cobalt hexaferrite nanoparticles using natural reducing agents (carbohydrates and pigments) by the sol-gel method and compared photocatalytic activities with their carbon-based nanocomposites including graphene and carbon nanotube. It was found that the degradation order of methylene orange under UV irradiation is arranged as graphene-based nanocomposite > CNT-based nanocomposite > pure nanoparticles. It may be due to the unique structure and a specific surface area of graphene (2D) and maybe CNT (1D). Kinetics studies showed that photocatalytic reactions in the presence of pure nanoparticles and CNT-based nanocomposites followed by pseudo-first-order, but in presence of graphene-based nanocomposites followed by zero-order indicates saturation of graphene surface with MO as an organic pollutant through physical absorption [54]. Till date, very few works has been done related to phytochemical-mediated synthesis of metal-/metal oxide-bounded graphene oxide/reduced graphene oxide nanostructures for the degradation of organic pollutants.

6. Photocatalytic degradation of pharmaceutical and personal care products

During the past few decades, the focus of environmental research has been more toward nonbiodegradable pharmaceuticals and personal care products because of its possible direct damages to living organisms and agriculture. PPCP can be further classified as antibiotics, anticonvulsants, contrast agents, hormones, nonsteroidal anti-inflammatory drugs (NSAIDs), β-blockers, lipid regulators, painkillers, preservatives, disinfectants, insect repellants, fungicides, soaps and detergents, fragrances, and sunscreen UV filters [55]. **Figure 5** shows the structural formula for different pharmaceutical and personal care products. Adsorption, photocatalysis, and a combination of these two are considered to be promising technologies for the removal of antibiotics from wastewater. The adsorption method has the advantages of easy operation, low cost, high efficiency, strong reproducibility, and availability of different adsorbents. Whereas photocatalysis, as the best advanced oxidation process (AOPs), is an economic, efficient, and green technology for degrading antibiotics using sunlight and ambient conditions [56]. Although graphene-based photocatalyst have the high potential for the photocatalytic degradation of antibiotics due to the interaction of the aromatic ring of antibiotics with graphene oxide/reduced graphene oxide mainly via π-π bond interaction.

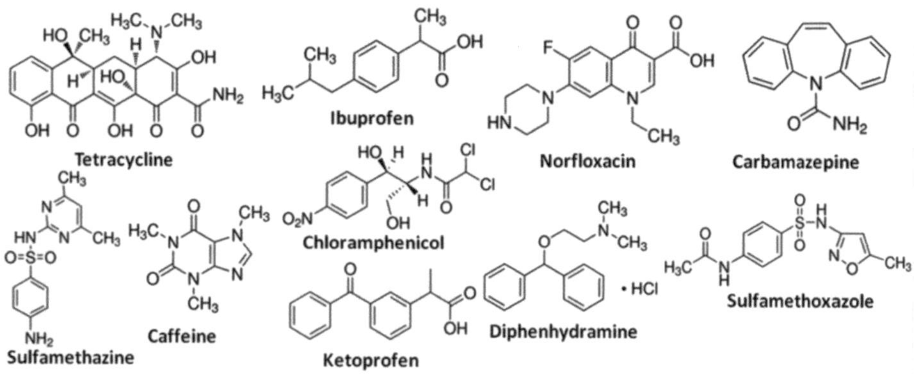

Figure 5.
The structural formula for different pharmaceutical and personal care products.

Tetracycline, the second most widely used antibiotics worldwide, has been extensively used as human medicine, as a veterinary drug, and as a growth promoter in animal cultivation. It is poorly metabolized and absorbed by humans and animals; thus large fractions are excreted via urine and feces. Therefore, it has been frequently detected in soil, surface waters, and even drinking water. Looking at this issue, Song et al. prepared MnO_2/graphene nanocomposite by an in situ hydrothermal method, and it successfully removed up to 99.4% of the tetracycline residue in pharmaceutical wastewater [57]. On the other hand, Shanavas et al. studied the degradation of ibuprofen and tetracycline molecules efficiently under visible light irradiation within 90 min using ternary $Cu/Bi_2Ti_2O_7$/rGO composite. The analysis of the obtained results suggested that the Cu nanoparticles and the rGO sheets play a major role in the photocatalytic ability of $Cu/Bi_2Ti_2O_7$/rGO photocatalysts by acting as charge carrier trappers and the suppression of e^--h^+ pair recombination [58]. A novel nitrogen-modified reduced graphene oxide nanocomposite with the function of adsorption and catalytic degradation of norfloxacin and ketoprofen in water was successfully explained by Peng et al. A complete removal of norfloxacin (20 mg/L) was achieved within 210 min with the addition of 10 mmol/L $S_2O_8^{2-}$ and 100 mg/L of as-prepared nanocomposite, and the removal efficiency of total organic carbon (TOC) was 89% [59].

The enhanced photocatalytic activities of graphene-TiO_2 composites make them an ideal material for remediating PPCP. However, one of the major obstacles to using nanoscale catalysts in water treatment is the difficulty in separating and recovering the catalysts. Lin et al. synthesized a series of TiO_2-reduced graphene oxide-coated side-glowing optical fibers (SOFs) by polymer-assisted hydrothermal deposition method. Photocatalytic performance of the synthesized nanocomposites exhibited significantly higher photocatalytic activities than pure TiO_2. The highest catalytic activity was observed by 2.7% reduced graphene oxide, resulting in 54% degradation for carbamazepine, 81% for ibuprofen, and 92% for sulfamethoxazole after 180 min high-pressure UV irradiation; the mineralization rates of the pharmaceuticals were similar between 52 and 59% [60]. Pastrana-Martínez et al. synthesized a graphene oxide-modified TiO_2 composite through liquid phase deposition and employed it to photocatalytically degrade diphenhydramine under ultraviolet and visible light irradiation. It was noted that the total degradation and significant mineralization of diphenhydramine pollutant (in less than 60 min) was achieved under near-UV/Vis irradiation for the optimum 3.3–4.0% wt GO in as-synthesized composites [61]. 3D porous rGO-TiO_2 aerogel was shown to remove carbamazepine (10 ppm) by more than 99% within 90 min in aqueous solution. The macroporous 3D structure of the aerogel resulted in abundant surface sites, effective charge separation, improved mass transport of contaminants, and easy separation [62].

A family of titanium-derived cobalt nanoparticles (Co_3O_4/TiO_2) and amine-functionalized titania Co_3O_4/TiO_2 graphene oxide nanocomposite synthesized via sol-gel and hydrothermal routes exhibit excellent performance for the photocatalytic degradation of oxytetracycline under solar and visible irradiation. Heterojunction formation between a low concentration of discrete Co_3O_4 nanoparticles and anatase titania strongly promoted the photocatalytic oxidative degradation of oxytetracycline, which was further enhanced upon trace GO addition. Amine functionalized 2% wt Co_3O_4/TiO_2/GO exhibits excellent rates and stability toward oxytetracycline photodegradation under visible light irradiation [63]. Fe_3O_4/Mn_3O_4-rGO nanocomposite was also used for a comprehensive photocatalytic degradation of the aqueous sulfamethazine solution. The results revealed 99% sulfamethazine degradation efficiency at optimum conditions of 0.07 mm/L sulfamethazine concentration, 0.5 g/L of Fe_3O_4/Mn_3O_4-rGO nanocomposites, 35°C, pH 3, and hydrogen peroxide concentration of 6 mM [64]. Karthik et al. fabricated a nanocomposite consisting of graphene oxide-decorated cerium molybdate nanocubes (Ce $(MoO_4)_2/GO$)) and utilized it in the photocatalytic degradation of chloramphenicol under visible light irradiation. The Ce $(MoO_4)_2/GO$ composite displayed excellent photodegradation potential against the drug and showed higher degradation efficiency (99% within 50 min) than the pure Ce $(MoO_4)_2$ nanocubes. The impressive performance of the composites was assigned to the excellent separation of the photogenerated electrons and holes [65]. A magnetically recyclable GO-TiO_2 composite showed up to 99% removal of carbamazepine and caffeine within 60 min under UV irradiation, and this composite was fully recoverable and reusable by magnetic separation. The added benefit is that the GO-TiO_2 composite is fully recoverable, reusable, and easy to produce [66]. The electrospun one-dimensional graphene oxide-nanofiber TiO_2 composite was prepared by using polyvinylpyrrolidone (PVP) as a fiberizing carrier. It showed a higher photodegradation rate of 4-chlorophenol under visible light irradiation relative to pristine TiO_2 due to the enhanced separation efficiency of photogenerated electron-hole pairs [67].

7. Photocatalytic degradation of pesticides

Pesticides are naturally derived or synthetically produced chemicals meant to control and destroy the pests and weeds. It includes bactericides, fungicides, herbicides, and insecticides. The excessive use of pesticides comes with various environmental apprehensions and harms living organisms due to their carcinogenic effects. So, there is an urgent need to develop a technology which can detect and degrade these hazardous pesticides efficiently without harming the environment with low cost. Nanoscience and nanotechnology is the emerging field which can serve this purpose [68]. Graphene-based composite materials hold great potential as adsorbent and photocatalyst for pesticide remediation from wastewater. A strong π-π interaction of the organic contaminant with the aromatic ring of graphene is the mechanism behind the strong adsorption behavior of GO toward different pesticides [68]. **Figure 6** shows the list of some pesticides normally used in the agricultural field. Boruah et al. prepared nanocomposite with Fe_3O_4 NPs and reduced GO (rGO) for the removal of triazine pesticides. The electrostatic interaction between the nanocomposite and the pesticide helped in efficient adsorption and removal of the pesticide. The nanocomposite exhibited 93.61% adsorption efficiency for the pesticides. Easy recovery of these magnetic nanocomposites from the reaction mixtures by applying an external magnetic field was an added advantage

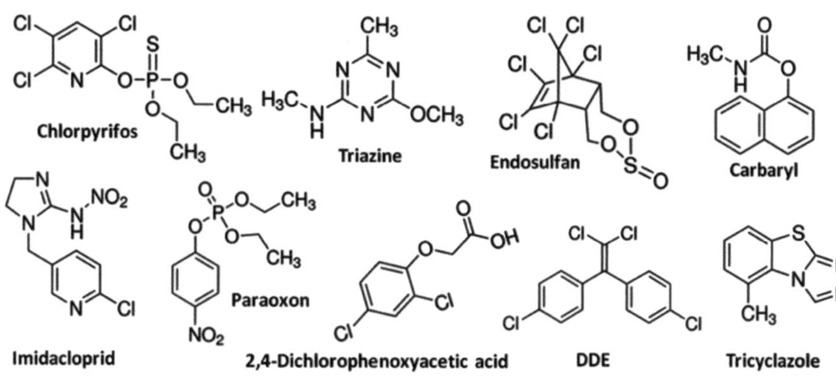

Figure 6.
The structural formula for different pesticides.

over their large and specific surface area [69]. The graphene-coated silica (GCS) nanocomposite was employed as a highly efficient adsorbent for removal of several organophosphorus pesticides from the contaminated water. The efficient adsorption was attributed to the electron-donating effects of N, S, and P atoms and the strong π-bonding network of benzene rings in the pesticides [70]. In another study, nanocomposite of rGO and silver nanoparticles has been used for the degradation and removal of organochlorine pesticides including dichlorodiphenyldichloroethylene (DDE), chlorpyrifos, and endosulfan. Silver nanoparticle-induced dehalogenation of pesticides, followed by adsorption of the degraded products on rGO, was the basic two-step mechanism behind the degradation and removal of these pesticides. The high reusability and adsorption capacity exhibited by this nanocomposite support its utilization for the remediation of different pesticides present in water and soil [71]. Cata et al. established a simple and cost-effective method of dispersing 15 nm silver nanoparticles on the surface rGO nanosheets uniformly using a one-pot hydrothermal process with the presence of polyvinylpyrrolidone as both the surfactant and reducing agent. The results indicated that the as-prepared rGO-Ag nanocomposites exhibit the highest surface-enhanced Raman spectroscopy (SERS) efficiency and good sensitivity with tricyclazole pesticide. These can be explained by the electron interactions between silver nanoparticles and graphene oxide, which make unique material advantages for the highly effective detection of pollutant molecules in the environmental monitoring applications [72].

Gupta et al. proposed the use of $CoFe_2O_4$@ TiO_2-rGO nanocomposites for the removal of an organophosphate insecticide, chlorpyrifos, from wastewater. The photocatalytic degradation of chlorpyrifos followed the pseudo-first-order kinetic model. The $CoFe_2O_4$ magnetic nanoparticles have already been proven as efficient photocatalysts for the degradation of environmental pollutants and could be easily separated by applying the external magnetic field [73]. Keihan et al. demonstrated the first time photodegradation of organophosphorus pesticide, paraoxon, using a nanocomposite. They prepared Ag nanoparticle and graphene co-loaded TiO_2 with various contents of Ag and graphene via a facile surfactant-free solvothermal method with a mixture of water and ethanol. It showed the best photocatalytic activity for the degradation of paraoxon containing nanocomposite with 6% wt Ag and 1% wt graphene under visible light irradiation. Further, complete photodegradation of paraoxon is confirmed by gas chromatography–mass spectrometry, and it produces 4-nitrophenol, diethylphosphate, mono-ethylphosphate, hydroquinone, and hydroxyhydroquinone as major intermediates, and subsequent photodegradation of these results in complete mineralization of paraoxon [74]. In another study, rGO has been utilized to fabricate nanocomposite with β-cyclodextrin/iron oxide for the removal of organochlorine pesticides from honey. The mechanism behind the removal of pesticide involved its adsorption on the nanocomposite based on magnetic solid-phase extraction. The prepared nanocomposite helped to detect very low concentrations of pesticides up to ppt and sub-ppt levels [75]. Improper discharge of chemical pesticides and herbicides into water bodies causes harmful effects on both the environment and human health. Ebrahimi et al. investigated the photocatalytic degradation of 2, 4-dichlorophenoxyacetic acid (2,4-D) using Mn-doped zinc oxide/graphene nanocomposite under light-emitting diode (LED) radiation in aqueous media. The results showed that 66.2% of 2,4-D could be photocatalytically degraded using Mn-doped zinc oxide/graphene nanocomposite under LED radiation at optimal conditions (pH 5, initial Zn concentration of 10 mg L^{-1}, nanocomposite concentration of 2 g L^{-1}, contact time of 120 min) [76]. Graphene oxide-titanium dioxide (GO@TiO_2) nanocomposite with mean diameter size of 14 nm has been fabricated by El-Shafai et al. and further used as photocatalyst for enhanced degradation of carbaryl and imidacloprid. It was found that the photocatalytic degradation efficiency of carbaryl and imidacloprid was 22 and 92.6% in the presence of GO@TiO_2 whereas 7 and 56.6% in the presence of TiO_2 alone. It suggests the superiority of GO@TiO_2 nanocomposite over TiO_2NPs for degradation of the examined toxic insecticides [77].

8. Conclusion

In conclusion, graphene oxide- and graphene-bounded metal nanocomposites have become more and more employed for wastewater treatment purposes and allow the design of next-generation photocatalyst systems. These nanocomposites may be good alternatives to the other materials such as TiO_2, CNT and fullerenes, or supported matrix. We reviewed the current advances in the fabrication and applications of the new nanocomposite materials for the remediation of toxic organic pollutants, dyes, pharmaceutical and personal care products, and pesticides. It also suggested the low cost, simple, and ecofriendly technique development to solve the pure water crisis in the coming years. The most common systems reported in the literature are based on the dispersion of metallic nanoparticles on the surface of the GO/rGO. As we know, the properties of photocatalysts are highly dependent on the surface structure of materials. On looking the benefits of plant materials for the synthesis of nanoparticles, various plant materials are also reported for in situ reduction of graphene oxide and metallic salts to fabricate graphene-bounded metallic nanocomposite, efficiently. It showed the role of GO/rGO as the sensitizer to metallic nanoparticles and enhances the catalytic effect in solar light. In this regard, this book chapter highlights the importance environment-friendly GO-/rGO-bounded metal nanocomposite and renewable energy for the environmental protection by following the concept "nature purification by natural materials."

Acknowledgements

This scientific work has been funded by the TATA College, Kolhan University, Chaibasa, Jharkhand, India.

Author details

Brajesh Kumar
Post Graduate Department of Chemistry, TATA College, Kolhan University, Chaibasa, Jharkhand, India

*Address all correspondence to: krmbraj@gmail.com; krmbrajnano@gmail.com

References

[1] Feynman RP. There's plenty of room at the bottom. Caltech's Engineering & Science Magazine. 1960;**23**(5):22-36

[2] Heiligtag FJ, Niederberg M. The fascinating world of nanoparticle research. Materials Today. 2013;**16**:262-271

[3] Vatta LL, Sanderson RD, Koch KR. Magnetic nanoparticles: Properties and potential applications. Pure and Applied Chemistry. 2006;**78**:1793-1801

[4] Kumar B, Vizuete KS, Sharma V, Debut A, Cumbal L. Ecofriendly synthesis of monodispersed silver nanoparticles using Andean Mortiño berry as reductant and its photocatalytic activity. Vacuum. 2019;**160**:272-278

[5] Kumar B, Smita K, Cumbal L. Biosynthesis of silver nanoparticles using lavender leaf and their applications for catalytic, sensing and antioxidant activities. Nanotechnology Reviews. 2016;**5**(6):521-528

[6] Mauter MS, Elimelech M. Environmental applications of carbon-based nanomaterials. Environmental Science & Technology. 2008;**42**:5843-5859

[7] Iravani S, Varma RS. Green synthesis, biomedical and biotechnological applications of carbon and graphene quantum dots. A review. Environmental Chemistry Letters. 2020;**18**:703-727

[8] Boehm H, Setton R, Stumpp E. Nomenclature and terminology of graphite intercalation compounds. Carbon. 1986;**24**:241-245

[9] Castro Neto AH, Guinea F, Peres NMR, Novoselov KS, Geim AK. The electronic properties of graphene. Reviews of Modern Physics. 2009;**81**(1):109-162

[10] Kumar A, Lee CH. Synthesis and biomedical applications of graphene: Present and future trends. In: Aliofkhazraei M, editor. Advances in Graphene Science. London: IntechOpen; 31st July 2013. DOI: 10.5772/55728. Available from: https://www.intechopen.com/books/advances-in-graphene-science/synthesis-and-biomedical-applications-of-graphene-present-and-future-trends

[11] Zhu Y, Murali S, Cai W, Li X, Suk W, Potts J, et al. Graphene and graphene oxide: Synthesis, properties and applications. Advanced Materials. 2010;**22**:3906-3924

[12] Li J, Dodgen L, Ye Q, Gan J. Degradation kinetics and metabolites of carbamazepine in soil. Environmental Science & Technology. 2013;**47**:3678-3684

[13] Li C, Wang X, Chen F, Zhang X, Wang K, Cui D. The antifungal activity of graphene oxide-silver nanocomposites. Biomaterials. 2013;**34**:3882-3890

[14] Martínez C, Canle M, Fernández M, Santaballa J, Faria J. Aqueous degradation of diclofenac by heterogeneous photocatalysis using nanostructured materials. Applied Catalysis B: Environmental. 2011;**107**:110-118

[15] Kamat P. Photophysical, photochemical and photocatalytic aspects of metal nanoparticles. The Journal of Physical Chemistry B. 2002;**106**:7729-7744

[16] Ali I, Basheer AA, Mbianda XY, Burakov A, Galunin E, Burakova I, et al. Graphene based adsorbents for remediation of noxious pollutants from wastewater. Environment International. 2019;**127**:160-180

[17] Xion Z, Zhang L, Ma J, Zhao X. Photocatalytic degradation of dyes over graphene-gold nanocomposites under visible light irradiation. Royal Society of Chemistry. 2010;**46**:6099-6101

[18] Ai L, Zhang C, Chen Z. Removal of methylene blue from aqueous solution by a solvothermal-synthesized graphene/magnetite composite. Journal of Hazardous Materials. 2011;**192**:1515-1524

[19] Lerf A, He H, Forster M, Klinowski J. Structure of graphite oxide revisited. The Journal of Physical Chemistry. B. 1998;**102**:4477-4482

[20] Machado BF, Serp P. Graphene-based materials for catalysis. Catalysis Science & Technology. 2012;**2**:54-75

[21] Rahaman O, Chellasamy V, Ponpandian N, Amirthapandian S, Panigrahi B, Thangadurai P. A facile green synthesis of reduced graphene oxide by using pollen grains of *Peltophorum pterocarpum* and study of its electrochemical behavior. Royal Society of Chemistry. 2014;**4**:56910-56917

[22] Maddinedi S, Kumar B, Vankayala R, Kaluru P, Pamanji S. Bioinspired reduced graphene oxide nanosheets using *Terminalia chebula* seeds extracts. Spectrochimica Acta Part A. 2015;**145**:117-124

[23] Huang X, Ying Z, Wu S, Qi X, He Q, Zhang Q, et al. Graphene-based materials: Synthesis, characterization, properties, and applications. Small. 2011;**14**:1876-1902

[24] Shi P, Su R, Wan F, Zhu M, Li D, Xu S. Co_3O_4 nanocrystals on graphene oxide as a synergistic catalyst for degradation of Orange II in water by advanced oxidation technology based on sulfate radicals. Applied Catalysis B: Environmental. 2012;**123-124**:265-272

[25] Cheviron P, Gouanvé F, Espuche E. Green synthesis of colloid silver nanoparticles and resulting biodegradable starch/silver nanocomposites. Carbohydrate Polymers. 2014;**108**:291-298

[26] Iravani S. Green synthesis of metal nanoparticles using plants. Green Chemistry. 2011;**13**:2638-2650

[27] Sharma V, Yngard R, Lin Y. Silver nanoparticles: Green synthesis and their antimicrobial activity. Advances in Colloid and Interface Science. 2009;**145**:83-96

[28] Zhang X, Chen YL, Liu R-S, Tsai DP. Plasmonic photocatalysis. Reports on Progress in Physics. 2013;**76**:046401

[29] Schlölg R, Hamid S. Nanocatalysis: Mature science revisited or something really new? Angewandte Chemie, International Edition. 2004;**43**:1628-1637

[30] Christopher P, Ingram DB, Linic S. Enhancing photochemical activity of semiconductor nanoparticles with optically active Ag nanostructures: Photochemistry mediated by Ag surface plasmons. Journal of Physical Chemistry C. 2010;**114**(19):9173-9177

[31] Zhang LL, Xiong Z, Zhao XS. Pillaring chemically exfoliated graphene oxide with carbon nanotubes for photocatalytic degradation of dyes under visible light irradiation. ACS Nano. 2010;**4**(11):7030-7036

[32] Williams G, Seger B, Kamat PV. TiO_2-graphene nanocomposites. UV-assisted photocatalytic reduction of graphene oxide. ACS Nano. 2008;**2**(7):1487-1491

[33] Kemp KC, Seema H, Saleh M, Le NH, Mahesh K, Chandra V, et al. Environmental applications using graphene composites: Water remediation and gas adsorption. Nanoscale. 2013;**5**:3149-3171

[34] Vizuete KS, Kumar B, Vaca AV, Debut A, Cumbal L. Mortiño (*Vaccinium floribundum* Kunth) berry assisted green synthesis and photocatalytic performance of silver–graphene nanocomposite. Journal of Photochemistry and Photobiology, A: Chemistry. 2016;**329**:273-279

[35] Chandu B, Nurbasha S, Bollikolla HB. A facile green reduction for graphene-silver nanocomposite using betel leaf extract for the photocatalytic degradation of water pollutants. ChemistrySelect. 2017;**2**:11172-11176

[36] Chandu B, Kurmarayuni CM, Kurapati S, Bollikolla HB. Green and economical synthesis of graphene–silver nanocomposite exhibiting excellent photocatalytic efficiency. Carbon Letters. 2020;**30**:225-233

[37] Maryami M, Nasrollahzadeh M, Mehdipour E, Sajadi SM. Preparation of the Ag/RGO nanocomposite by use of *Abutilon hirtum* leaf extract: A recoverable catalyst for the reduction of organic dyes in aqueous medium at room temperature. International Journal of Hydrogen Energy. 2016;**41**:21236-21245

[38] Chandra SP, Sourav Bag S, Bhar RB, Pramanik P. Mn_2O_3 decorated graphene nanosheet: An advanced material for the photocatalytic degradation of organic dyes. Materials Science and Engineering B. 2012;**177**:855-861

[39] Benjwal B, Kumar M, Chamoli P, Kar KK. Enhanced photocatalytic degradation of methylene blue and adsorption of arsenic(III) by reduced graphene oxide (rGO)–metal oxide (TiO_2/Fe_3O_4) based nanocomposites. RSC Advances. 2015;**5**:73249-73260

[40] Nenavathu BP, Kandula S, Verma S. Visible-light-driven photocatalytic degradation of safranin-T dye using functionalized graphene oxide nanosheet (FGS)/ZnO nanocomposites. RSC Advances. 2018;**8**:19659-19667

[41] Xiong Z, Zhang L, Zhao X. Visible-light-induced dye degradation over copper-modified reduced Graphene oxide. Chemistry—A European Journal. 2011;**17**:2428-2434

[42] Fu Y, Wang X. Magnetically separable $ZnFe_2O_4$–graphene catalyst and its high photocatalytic performance under visible light irradiation. Industrial and Engineering Chemistry Research. 2011;**50**:7210-7218

[43] Lui G, Liao JY, Duan A, et al. Graphene-wrapped hierarchical TiO_2 nanoflower composites with enhanced photocatalytic performance. Journal of Materials Chemistry A. 2013;**1**:12255-12262

[44] Lv T, Pan L, Liu X, Lu T, Zhu G, Sun Z. Enhanced photocatalytic degradation of methylene blue by ZnO-reduced graphene oxide composite synthesized via microwave-assisted reaction. Journal of Alloys and Compounds. 2011;**509**:10086-10091

[45] Thangavel S, Venugopal G, Jae KS. Enhanced photocatalytic efficacy of organic dyes using β-tin tungstate–reduced graphene oxide nanocomposites. Materials Chemistry and Physics. 2014;**145**:108-115

[46] Sharma K, Maiti K, Kim NH, Hui D, Lee JH. Green synthesis of glucose-reduced graphene oxide supported Ag-Cu_2O nanocomposites for the enhanced visible-light photocatalytic activity. Composites Part B: Engineering. 2018;**138**:35-44

[47] Xiong Z, Zhang LL, Ma J, Zhao XS. Photocatalytic degradation of dyes over graphene–gold nanocomposites under visible light irradiation. Chemical Communications. 2010;**46**:6099-6101

[48] Agorku ES, Mamo MA, Mamba BB, Pandey AC, Mishra AK.

Palladium-decorated zinc sulfide/reduced graphene oxide nanocomposites for enhanced visible light-driven photodegradation of indigo carmine. Materials Science in Semiconductor Processing. 2015;**33**:119-126

[49] Kurt BZ, Durmus Z, Durmus A. Preparation and characterization of platinum (Pt) and palladium (Pd) nanoparticle decorated graphene sheets and their utilization for the elimination of basic fuchsin and indigo carmine dyes. Solid State Sciences. 2016;**51**:51-58

[50] Isari AA, Payan A, Fattahi M, Jorfi S, Kakavandi B. Photocatalytic degradation of rhodamine B and real textile wastewater using Fe-doped TiO_2 anchored on reduced graphene oxide (Fe-TiO_2/rGO): Characterization and feasibility, mechanism and pathway studies. Applied Surface Science. 2018;**462**:549-564

[51] El-Shafai NM, El-Khouly ME, El-Kemary M, Ramadana MS, Masoud MS. Graphene oxide–metal oxide nanocomposites: Fabrication, characterization and removal of cationic rhodamine B dye. RSC Advances. 2018;**8**:13323-13332

[52] Wang L-L, He H-Y, Li Q, Ren Z-Q. Photocatalytic activities of SnS-reduced graphene oxide by the photodegradation of malachite green in water. Materials Research Innovations. 2016;**20**(6):458-464

[53] Ye A, Fan W, Zhang Q, Deng W, Wang Y. CdS–graphene and CdS–CNT nanocomposites as visible-light photocatalysts for hydrogen evolution and organic dye degradation. Catalysis Science & Technology. 2012;**2**:969-978

[54] Ansari F, Soofivand F, Salavati-Niasari M. Eco-friendly synthesis of cobalt hexaferrite and improvement of photocatalytic activity by preparation of carbonic-based nanocomposites for waste-water treatment. Composites Part B Engineering. 2019;**165**:500-509

[55] Yang Y, Ok YS, Kim K-H, Kwon EE, Tsang YF. Occurrences and removal of pharmaceuticals and personal care products (PPCPs) in drinking water and water/sewage treatment plants: A review. Science of the Total Environment. 2017;**596-597**:303-320

[56] Li MF, Liu Yun G, Zeng GM, Shao-bo Liu NL. Graphene and graphene-based nanocomposites used for antibiotics removal in water treatment: A review. Chemosphere. 2019;**226**:360-380

[57] Song Z, Ma Y-L, Li C-E. The residual tetracycline in pharmaceutical wastewater was effectively removed by using MnO_2/graphene nanocomposite. The Science of the Total Environment. 2019;**651**:580-590

[58] Shanavas S, Priyadharsan A, Gkanas EI, Acevedo R, Anbarasan PM. High efficient catalytic degradation of tetracycline and ibuprofen using visible light driven novel Cu/$Bi_2Ti_2O_7$/rGO nanocomposite: Kinetics, intermediates and mechanism. Journal of Industrial and Engineering Chemistry. 2019;**72**:512-528

[59] Peng G, Zhang M, Deng S, Shan D, He Q, Yu G. Adsorption and catalytic oxidation of pharmaceuticals by nitrogen-doped reduced graphene oxide/Fe_3O_4 nanocomposite. Chemical Engineering Journal. 2018;**341**:361-370

[60] Lin L, Wang H, Xu P. Immobilized TiO_2-reduced graphene oxide nanocomposites on optical fibers as high performance photocatalysts for degradation of pharmaceuticals. Chemical Engineering Journal. 2017;**310**:389-398

[61] Pastrana-Martínez LM, Morales-Torres S, Likodimos V, Figueiredo JL, Faria JL, Falaras P, et al. Advanced nanostructured photocatalysts based on reduced graphene oxide–TiO_2 composites for degradation of diphenhydramine pharmaceutical and

methyl orange dye. Applied Catalysis. B, Environmental. 2012, 2012;**123**:241-256

[62] Nawaz M, Miran W, Jang J, Lee DS. One-step hydrothermal synthesis of porous 3D reduced graphene oxide/ TiO_2 aerogel for carbamazepine photodegradation in aqueous solution. Applied Catalysis B: Environmental. 2017;**203**:85-95

[63] Jo W-K, Kumar S, Isaacs MA, Lee AF, Karthikeyan S. Cobalt promoted TiO_2/GO for the photocatalytic degradation of oxytetracycline and Congo red. Applied Catalysis B: Environmental. 2017;**201**:159-168

[64] Wan Z, Wang J. Degradation of sulfamethazine using Fe_3O_4-Mn_3O_4/ reduced graphene oxide hybrid as Fenton-like catalyst. Journal of Hazardous Materials. 2017;**324**:653-664

[65] Karthik R, Vinoth Kumar J, Chen S-M, Karuppiah C, Cheng Y-H, Muthuraj V. A study of electrocatalytic and photocatalytic activity of cerium molybdate nanocubes decorated graphene oxide for the sensing and degradation of antibiotic drug chloramphenicol. ACS Applied Materials & Interfaces. 2017;**9**:6547-6559

[66] Linley S, Liu Y, Ptacek CJ, Blowes DW, Gu FX. Recyclable graphene oxide-supported titanium dioxide photocatalysts with tunable properties. ACS Applied Materials & Interfaces. 2014;**6**:4658-4668

[67] Zhang L, Zhang Q, Xie H, Guo J, Lyu H, Li Y, et al. Electrospun titania nanofibers segregated by graphene oxide for improved visible light photocatalysis. Applied Catalysis B: Environmental. 2017;**201**:470-478

[68] Rawtani D, Khatri N, Tyagi S, Pandey G. Nanotechnology-based recent approaches for sensing and remediation of pesticides. Journal of Environmental Management. 2018;**206**:749-762

[69] Boruah PK, Sharma B, Hussain N, Das MR. Magnetically recoverable Fe_3O_4/graphene nanocomposite towards efficient removal of triazine pesticides from aqueous solution: Investigation of the adsorption phenomenon and specific ion effect. Chemosphere. 2016;**168**:1-10

[70] Liu X, Zhang H, Ma Y, Wu X, Meng L, Guo Y, et al. Graphene-coated silica as a highly efficient sorbent for residual organophosphorus pesticides in water. Journal of Materials Chemistry A. 2013;**1**:1875-1884

[71] Koushik D, Gupta SS, Maliyekkal SM, Pradeep T. Rapid dehalogenation of pesticides and organics at the interface of reduced graphene oxide-silver nanocomposite. Journal of Hazardous Materials. 2016;**308**:192-198

[72] Cata VV, Dinh NX, Tama LT, Quy NV, Phand VN, Le A-T. One-pot hydrothermal-synthesized rGO-Ag nanocomposite as a sensing platform for detection and quantification of methylene blue organic dye and tricyclazole pesticide. Materials Today Communications. 2019;**21**:100639

[73] Gupta VK, Eren T, Atar N, Yola ML, Parlak C, Karimi-Maleh H. $CoFe_2O_4$@ TiO_2 decorated reduced graphene oxide nanocomposite for photocatalytic degradation of chlorpyrifos. Journal of Molecular Liquids. 2015;**208**:122-129

[74] Keihan AH, Hosseinzadeh R, Farhadian M, Kooshki H, Hosseinzadeh G. Solvothermal preparation of Ag nanoparticle and graphene co-loaded TiO_2 for the photocatalytic degradation of paraoxon pesticide under visible light irradiation. RSC Advances. 2016;**6**:83673-83687

[75] Mahpishanian S, Sereshti H. One-step green synthesis of β-cyclodextrin/ iron oxide-reduced graphene oxide nanocomposite with high

supramolecular recognition capability: Application for vortex-assisted magnetic solid phase extraction of organochlorine pesticides residue from honey samples. Journal of Chromatography. A. 2017;**1485**:32-43

[76] Ebrahimi R, Mohammadi M, Maleki A, et al. Photocatalytic degradation of 2,4-dichlorophenoxyacetic acid in aqueous solution using Mn-doped ZnO/Graphene nanocomposite under LED radiation. Journal of Inorganic and Organometallic Polymers. 2020;**30**: 923-934

[77] El-Shafai NM, El-Khouly ME, El-Kemary M, Ramadan MS, Derbalah AS, Masoud MS. Fabrication and characterization of graphene oxide–titanium dioxide nanocomposite for degradation of some toxic insecticides. Journal of Industrial and Engineering Chemistry. 2019;**69**:315-323

Chapter 4

Versatile and Scalable Approaches to Tune Carbon Black Characteristics for Boosting Adsorption and VOC Sensing Applications

Michela Alfè and Valentina Gargiulo

Abstract

This chapter explores the potential use of carbon black (CB) as point of departure to design a highly varied array of materials with practical applications in energetics, remediation, and sensoristics. Thanks to its graphenic moieties embedded in a nanostructured backbone, CB is prone to be structurally and chemically modified exploiting quite mild chemical conditions. The proposed approaches, implying an easy tuning of the chemo-physical properties to the specific needs, are thought up to meet urgent sustainability needs: low costs, scalability, and flexibility. In this chapter, we will describe the modification of CB at the surface (i.e., introduction of oxygen functional group, functionalization, coating with active phases) and a highly CB destructuration to produce graphene-related materials (GRMs) suitable for film production and for the designing of new hybrid materials. CB is converted into highly homogenous CB-modified nanoparticles (around 160 nm) with adjustable surface properties (hydrophilicity, type and surface charge density, pore size distribution) and in highly versatile GRMs for the production of structured electrical conductive ultrathin films for trace alcohol sensing and a wide array of hybrid materials, including photocatalysts (carbon-iron oxide, silica-carbon, carbon-titanium oxide hybrids) for adsorption applications (CO_2 capture, heavy metal capture).

Keywords: carbon black, advanced materials, graphene-related materials, ultrathin films, hybrid materials, sensing layers

1. Introduction

Carbons blacks (CB) are carbonaceous materials [1] broadly classified as industrial aciniform aggregates (IAA) in which the smallest dispersible unit consists of grapelike aggregates. They are mostly produced from controlled thermal processes mainly starting from crude oils or natural gas. The different manufacture processes designate the different kinds of CB, each of them characterized by specific morphologies and chemo-physical properties [1]. The most widely used production processes, providing CB with a highly controlled morphology, are oil furnace (also known as furnace black process) and thermal black processes in which heavy aromatic oils and natural gas (methane is the mostly used feedstock) or higher-grade hydrocarbon oils are pyrolyzed (furnace CB) and thermally decomposed (thermal CB), respectively. CB are also produced by channel black process (providing crude oil-generated CB pigments with very fine particles characterized by a high content of oxygen-containing surface groups), lamp black process (the oldest way to produce CB, not suitable for mass production, providing CB with specific color from oil lamp or pinewood heating), and acetylene black process (providing a high-purity, high-crystalline, and electric conductive CB by thermally decomposing acetylene gas) [2, 3].

CB are composed almost exclusively by elemental carbon (>97 wt.%) with a minor and variable amount of hydrogen, oxygen, and sulfur (less than 1% each); usually the hydrogen to carbon atomic ratio (H/C) is less than 0.05 [2, 4]. Trace amounts of adsorbed organic compounds (polycyclic aromatic hydrocarbons and aliphatic/aromatic hydrocarbons) and inorganic impurities (salts of alkali and alkaline earth metals) as manufacturing residuals can also be found [4].

At the nanoscale level, CB is composed almost exclusively of stacked sheets of condensed aromatic ring systems (i.e., graphene layers), concentrically arranged and oriented around a central core to form almost spherical primary particles (from 10 to 300 nm, in dependence of the manufacturing). Primary particles are fused into grapelike-shaped aggregates (aciniform aggregates, usually less than 500 nm) representing the discrete, dispersible CB smallest unit. Aggregates clustered in the so-called agglomerates.

CB are usually highly agglomerated, with 10–1000 aggregates per agglomerate. The three-dimensional configuration of the CB aggregates results in varied but usually large surface areas (SA) (5–200 m^2/g).

The size of the primary particle and the degree of aggregation differentiate the grades of carbon black [2]. The CB microstructure (microtexture, particle size, and the morphology of aggregates) is regarded as the key property to allocate the proper commercial use of a specific carbon black [2].

The edge of CB graphenic layers is chemically reactive and can be decorated with specific functional groups, while the basal planes are quite less reactive. Several evidences have confirmed the presence of H atoms as well as four types of oxygen-containing functional groups on the as-produced CB surface: lactones, phenolic, carboxylic, and carbonylic groups [5, 6]. Such chemical groups are considered to be essential for the chemical manipulation of CB powder in mild conditions since they modulate CB solvent suspendibility.

Furnace black is the most commonly used CB (almost 95% of the total market) as it is widely applied in tire and other related automotive manufacture, while only some specific typologies of thermal black find application in rubber production. Although CB are massively used in ink and rubber industry, noteworthy emerging CB applications are the production of sensitive components in electrochemical sensors [7], the production of counter electrodes for dye-sensitized solar cells [8], the production of hybrid photocatalysts [9] and composites with heat insulating properties [10], the development of electrically conductive fabrics for wearable heating applications [11], and the development of construction materials with improved physical properties (i.e., electrical conductivity) [12, 13].

In our approach, we explored the production of a highly varied array of advanced materials starting from a furnace CB and using feasible approaches designed to meet the continuous need of low-cost and scalable protocols for tuning easily the relevant chemo-physical properties of the desired carbon-based materials.

An extensive experimental campaign has been conducted since many years in our group on a CB N110 type (furnace black, **Figure 1**).

This CB is characterized by a H/C atomic ratio of 0.058, a density of 1.8 g/mL at 25°C, and an inorganic content less than 0.1 wt.%. It is a meso-porous material

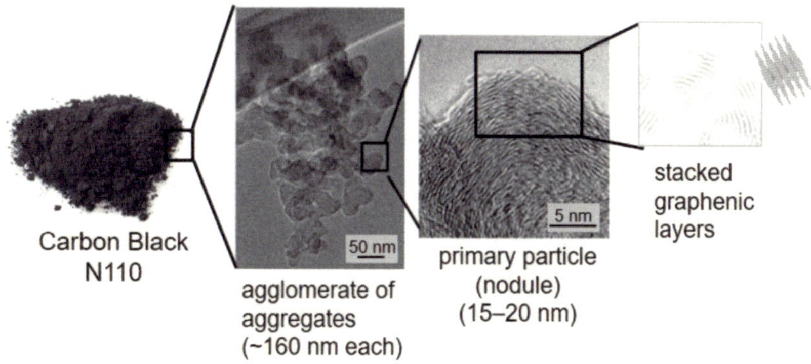

Figure 1.
Structure of a CB N110 (furnace black) from macro- to nanoscale.

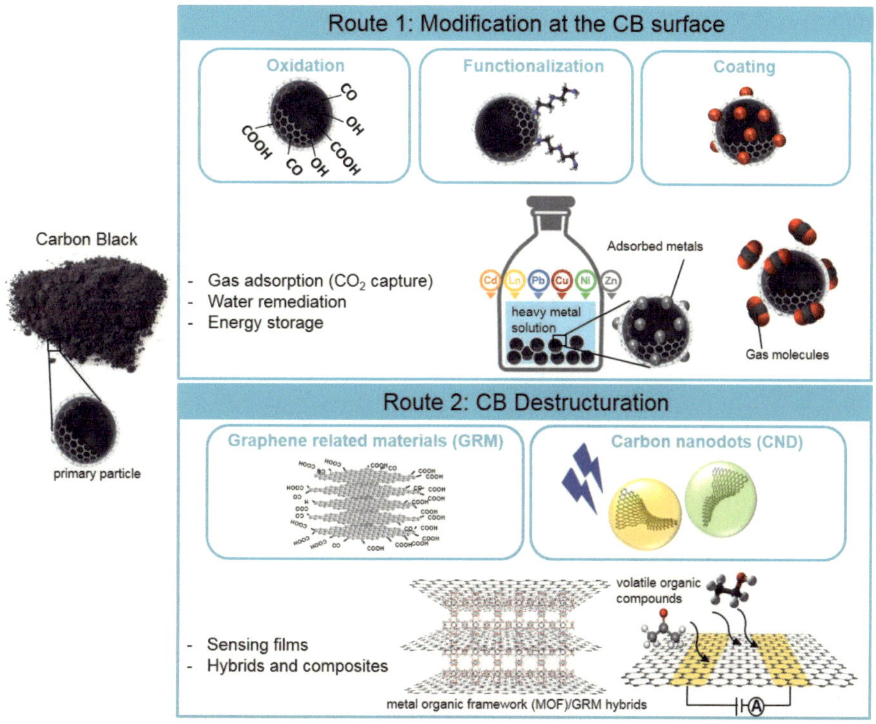

Figure 2.
Advanced materials produced from CB N110 and their applications in sensing and remediation fields.

with a SA of 139 m^2/g. Its microstructure is organized in chain-like aggregates with a hydrodynamic diameter, measured by dynamic light scattering (DLS), of 160 ± 20 nm. The diameter of the aggregate building blocks (primary particles or nodules) is 15–20 nm [14, 15].

Two main routes have been exploited starting from the as-produced CB N110 (hereinafter CB) (**Figure 2**) [16]:

i. Modification and introduction of specific chemical functional groups at the CB surface (modification at the CB surface).

ii. Extraction of graphenic nanoplatelets from CB backbone (CB destructuration).

The first route allows to produce CB-modified nanoparticles having narrow hydrodynamic diameter (160 ± 10 nm) and customized surface properties (hydrophilicity, type and surface charge density, SA, and pore size distribution). These CB-derived materials have been mainly designed as adsorbents for heavy metal removal (thanks to the presence of oxygen functional groups) and as supports for ionic liquid (SILP) and for magnetic iron oxide particles (thanks to their meso-porosity).

The second route is framed as a top-down approach and leads to the demolition of CB backbone to obtain water-stable graphene-related materials (GRM) as (i) stacked graphene-like layers (GL), highly functionalized to serve as building blocks for producing a wide array of advanced hybrid materials and ultrathin self-assembling carbon-based films, and (ii) yellow and green fluorescent carbon nanodots (the latter at the early stage of the experimentation).

In the following, details on the synthetic approaches, the structural characteristics, and potential applications of the developed materials in environmental fields (sensing and remediation) are reported.

2. Modification of CB at the surface

2.1 Colloidal hydrophilic nanoparticles (carboxyl enriched)

The oxidative treatment of CB with hot nitric acid allows to produce colloidal hydrophilic nanoparticles (HNPs) with uniform size bearing on the surface oxygen-containing acidic sites (mainly carboxylic groups but also lactone, lactol, and phenol groups) [6, 15, 17] whose amount depends on the oxidative treatment time (the total number of oxygen-containing functional groups varies from 1.56 ± 0.10 mmol/g to 2.25 ± 0.08 mmol/g upon 4 and 24 h of treatment, respectively). The wet chemistry treatment, performed up to 24 h of reaction time, generates nanoparticles with an increasing oxygen content: from 0.59 wt.% in as-produced CB up to 31.8 wt.% after 24 h of reaction time [15]. A higher reaction time does not lead to a significant gain in oxygenation degree (further discussed in detail later on) as testified by the product obtained after 90 h of treatment characterized by an oxygen content of 44.1 wt.% [17].

Interestingly, the acid treatment up to 24 h does not alter the hydrodynamic diameter of the aggregate (around 165 ± 10 nm in all the explored cases) and significantly improves the volumes of both micro- and meso-pores [15] thus enhancing the surface area from 139 m^2/g (raw CB) up to 323 m^2/g (after 24 h of treatment).

The HNPs obtained after treatments lasting between 4 and 24 h exhibited a good colloidal stability in water and in a wide pH range (from 4 to 12) [6, 15] and were found highly exploitable for heavy metal removal including rare-earth elements (Ln(III) ions) [6, 15].

Their capacity to reversibly adsorb heavy metals was verified through batch experiments by using cadmium, nickel, lead, and zinc as target metals [6, 15, 17]. In the case of cadmium, an appreciable adsorption capacity at different temperatures (10–60°C) and in a wide range of pH (2–6) has been verified [15]. A good correlation between the concentration of active sites (specifically carboxylic functional groups) on the HNP surface and the cadmium adsorption capacity was also highlighted [15]. Interestingly enough, the HNPs exhibited good cadmium capture also at low concentration: this circumstance is noticeable since in water depuration the actual adsorption capacity close to the allowed discharge limit is a very important parameter. In particular, accordingly with the Italy's current regulations that fix the limit for sewage discharge at 0.2 mg/L [18], it was established that the cadmium adsorption capacity at 20°C and pH 3.5 (selected by taking into account that most of the cadmium wastewaters are acidic) was 3.84 mg/g for HNP at the higher oxygenation degree (24 h of hot acidic treatment) [15]. These values resulted comparable or superior with respect to commercial activated carbons (even those oxidized), ashes, and residual solids but similar or far lower than specific sorbents from biosources (activated char) or polyacrylonitrile (PAN) fibers [15].

The heavy metal sorption mechanism on HNPs has been also investigated by combining X-ray fluorescence, infrared spectroscopy, and differential pulse voltammetry at pH $\leq$ 6 [6]. The experiments suggested that the adsorption of metal ions on HNPs is controlled by chemical adsorption involving a strong complexation of metal ions with the carboxylic groups on the surface of HNPs.

Moreover, the good surface complexation of lanthanum ions on HNP surface highlighted by this study opens the possibility to feasibly exploit HNPs in solid-phase extraction for the pre-concentration, separation, and quantification of rare-earth elements, also including actinides [6].

As mentioned before, the HNPs are highly stable in water over a wide pH range given the establishment of colloidal stable suspensions, causing difficulties in the separation of HNPs from water. To overcome this limitation, micrometric-sized carbon-silica hybrids (SHNPs) have been designed, developed, and tested for heavy metal adsorption [15, 17]. The HNPs were firmly supported over silica thanks to the tight interactions between the oxygen functional groups borne by the HNP and the hydrophilic silica surface. The carbon-silica hybrids can be easily filtered from wastewater with conventional filter units, and, moreover, they can be used also in applications at pH values higher than those explored with bare HNPs [15], thus paving the way to a feasible use of these sorbents for heavy metal adsorption in commercial units [15, 17].

HNPs obtained after different times of oxidation (from 0.5 up to 90 h) have been used for the preparation of different SHNPs. The adsorption capacity of HNPs supported over silica toward Cd(II) (representative of a soft acid) and Ni(II) ions (representative of an intermediate acid) [19] has been determined through batch experiments exploring the effects of the most relevant process parameters (solution temperature, pH, and initial concentration of the target metal) [17]. The effect of different sorbent preparations (HNP oxidation time and loading of active phase over the silica support) was also thoroughly investigated [17]. The SHNPs exhibited high removal efficiencies for the three metals at neutral pH and low temperature (10°C), conditions typically adopted for natural water and for some industrial wastewater remediation (**Figure 3**).

As a consequence of the presence of the carboxylic functional groups on the HNP surface, the SHNPs resulted more selective toward Ni(II) ions, which behave as an intermediate acid, than toward Cd(II) ions, which behave as a soft acid. Among Cd(II) and Pb(II) ions, the latter exhibited the lowest adsorption efficiency [17].

The sorbents exhibited high adsorption capacities per gram of active phase: considering the silica-based material obtained with the HNPs produced after 24 h of oxidation, the following adsorption capacities have been measured: 13.48 mg/g for Ni(II) ions, which is far higher than those of other sorbents available in the literature, 0.54 mg/g for Cd(II) ions, and 8.87 mg/g for Pb(II) ions [17].

Modeling analysis on the adsorption isotherms revealed that the Gibbs free energy of interactions between the sorbent and Ni(II) and Pb(II) ions is higher

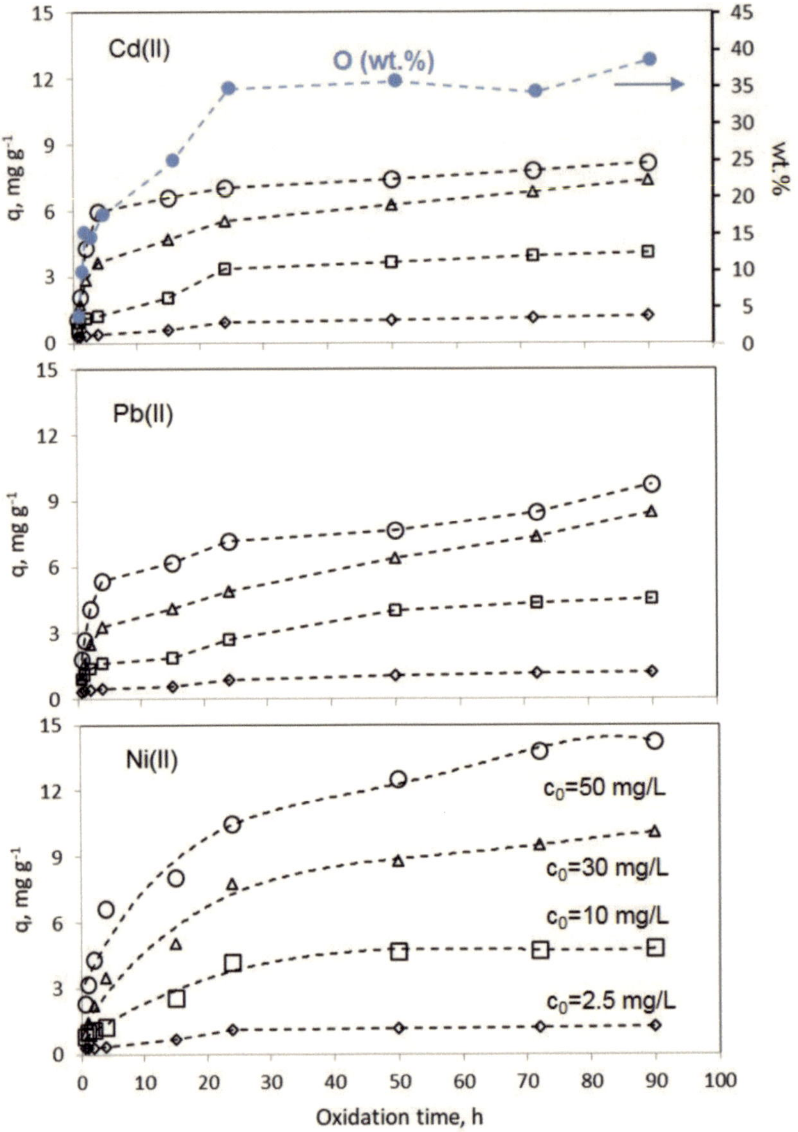

Figure 3.
Adsorption capacities of SHNPs toward Cd(II), Pb(II), and Ni(II) ions as a function of CB oxidation time. Experimental conditions: 5 wt.% HNP loading, T = 25°C, m/V = 40 g/L; ◊ C_o = 2.5 mg/L; □ C_o = 10 mg L/; Δ C_o = 30 mg/L; ○ C_o = 50 mg/L. The oxygen content (wt.%) of the corresponding HNPs is also reported (blue dots) (adapted from [17]).

than that of Cd(II) ions indicating that the sorbents are more affine to intermediate acids, as Ni(II) and Pb(II) ions, than to soft acids, as Cd(II) ions [17].

As concerns the effect of sorbent preparation parameters, it is worth of note that an active phase obtained after 90 h of oxidation with respect to a phase obtained after 24 h may provide an increase in the adsorption capacity of Ni(II) and Pb(II) ions by an additional 30%, while no other benefits have been sought for Cd(II) ions.

2.2 Surface-modified CB (carboxyl and amino-derivatives, CB/Fe_3O_4, supported ionic liquids)

Since 2005 the widespread carbon capture and storage (CCS) program aims to prevent the effects of carbon dioxide emissions. The CCS concept is a multistep and multidisciplinary process based on (i) catching CO_2 from massive sources, such as fossil fuel-based power plants; (ii) transportation to the designed place of storage; and (iii) stable storage where it will not get to the atmosphere, usually underground geological formations [20].

As concerns the first step, to date three methods of capturing CO_2 directly from industrial sources are known post-combustion capture, precombustion capture, and combustion of fossil fuels in a pure oxygen environment [20]. Samanta et al. indicate that post-combustion capture of CO_2 from flue gas is one of the key technological options to meet the need of reducing greenhouse gases in the short-term period, because the capture units can be potentially added without substantially modifying at all the existing fleet of coal-fired power plants [21]. The most performing post-combustion carbon capture processes include amine-based absorption, ammonia-based absorption, solid sorbent adsorption, and membrane filtration [20].

Adsorption processes using solid sorbents capable of capturing CO_2 from flue gas streams have shown many potential advantages, compared to other conventional CO_2 capture approaches as aqueous amine solvents.

The use of solid sorbents is one of the most promising options for post-combustion CO_2 capture strategies [21] since it assures (i) reduced energy consumption for regeneration, (ii) great capacity, (iii) selectivity, and (iv) ease of handling. In this framework, the specific characteristics required by a good CO_2 adsorbent are (i) production and regeneration at reduced (or, at least, economically convenient) costs; (ii) selectivity toward CO_2 over the typical flue gas components; and (iii) chemical and mechanical stability to undergo repeated CO_2 sorption/desorption cycles over a wide range of temperatures and pressures [22].

Overall, materials with a distinctive surface chemistry and defined porosity (pore diameter below 1 nm) find large applications in CCS technologies [21–24]. In the manifold of inorganic materials suitable for CO_2 capture applications, ferric oxide nanoparticles represent a cost-effective CO_2 sorbent, thanks to the presence of surface coordinatively unsaturated metal and O sites favoring the instaurations of acid–base interactions (Lewis type) with the acidic CO_2 molecule. The interaction between the active sites on iron oxide surface and CO_2 leads to the formation of adsorbed carbonates, bicarbonates and carboxylates, as well as bent CO_2 species [25–27]. The efficient use of ferric oxide as CO_2 adsorbent under post-combustion CO_2 capture conditions is hampered by the tendency to form agglomerates with lower surface area and lower gas capture efficiency. To limit this issue, the dispersion of ferric oxide nanoparticles over a carbonaceous matrix (activated carbons, carbon nanotubes, graphite oxide) is successfully adopted [26, 27]. CB has been proposed as carbonaceous support for iron oxide (specifically magnetite, Fe_3O_4, FM) to produce hybrid sorbents for CO_2 capture applications [28–30]. In this framework hybrid materials were synthetized applying a one-pot synthetic approach based on a coprecipitation strategy allowing the iron oxide particles to form in presence of CB [28, 29]. In this way a homogeneous incorporation of both the carbonaceous material and the iron oxide particles into the hybrid materials was obtained. A series of CB/FM hybrids was produced by varying the CB to FM ratio. The obtained materials exhibited a comparable SA (around 150 m^2/g) and tunable pore size distributions [28]. More specifically, the pore size distribution ranges from the porosity typical of FM (20–50 Å) for low-CB-containing hybrids (up to 14 wt.% of CB load) to the CB pore size distribution typically broadly centered around 200 Angstroms (for hybrid containing CB more than 60 wt.% of CB) [28]. The CO_2 uptakes of the different hybrids were evaluated in a lab-scale fixed bed reactor under dynamic conditions using typical post-combustion conditions (CO_2 3–10% vol. and atmospheric pressure). It was established that the CO_2 uptake reached the highest values when the amount of CB in the hybrid ranged between 14.3 and 60 wt.%. This finding indicated that a synergistic effect between the two components (CB and FM) has been reached. It was also demonstrated that both the material surface area (accounting for physisorption phenomena) and surface chemical features of iron oxide particles (accounting for chemisorption phenomena) have relevance [28, 30].

In order to have a clear idea of the potential of the CB/FM hybrids, the sorbent exhibiting the highest adsorption capacity (50 wt.% of CB load, 16.5 mg_{CO2}/g at $C_0(CO_2)$ = 3% vol., and inlet flow rate = 15 Nl/h) was tested in a prototypal sound-assisted fluidized bed [28]. The fluidization technology was adopted for this testing since it is generally considered to be one of the best available techniques to handle and process large quantities of powders. Since the fluidization of fine powders (as CB/FM hybrids, since the particle size distribution is lower than 40 μm with a Sauter diameter of about 350 nm) is expected to be particularly difficult, sound-assisted fluidization was used to achieve a smooth fluidization regime by overcoming the interparticle forces.

The testing of the selected sorbent was carried out operating the sound-assisted fluidized bed apparatus under post-combustion conditions ($C_0(CO_2)$ = 10% vol.; inlet flow rate = 67.8 Nl/h; sound intensity = 140 dB; sound frequency = 80 Hz) also exploring in situ regeneration and cyclability (desorption temperature = 250°C). In such conditions the CO_2 uptake was considerably enhanced up to about 20 mg_{CO2}/g. The experimental campaign demonstrated also that the material underwent several CO_2 adsorption and desorption cycles without losing in performances. It is noteworthy that the time needed to adsorb 95% of the total CO_2 uptake (about 11 min) is comparable to the required to completely regenerate the sorbent, opening the possibility to operate cyclically in two interconnected fluidized beds [28].

The best performing hybrid (50 wt.% of CB load) was fully characterized from both thermodynamic and kinetic points of view under dynamic conditions [31, 32]. More in detail, the CO_2 adsorption data collected under isothermal conditions (18–150°C) and with CO_2 partial pressure ranging between 0.5 and 20 vol.% [31] showed that adsorption temperature and pressure have opposite effects on the thermodynamics of the process: the CO_2 adsorption capacity increases with increasing pressure, and it decreases with increasing temperature. The Freundlich and Toth models allowed the best fits of the CO_2 adsorption data suggesting that a multilayer process characterized by a heterogeneous surface binding takes place. Finally, the CO_2 adsorption on the selected CB/FM hybrid was found to be a spontaneous and exothermic process based neither on purely physisorption nor on purely chemisorption [31]. From the kinetic point of view, the mechanism of CO_2 adsorption on the selected CB/FM hybrid has been investigated in terms of controlling mass transfer resistances, interparticle diffusion, intraparticle diffusion, and Boyd's film-diffusion models [32]. The results showed that mass transfer during CO_2 adsorption on CB/FM proceeds through a diffusion-based process involving three different successive steps: film diffusion, intraparticle diffusion, and surface reaction once equilibrium is approaching. The adsorption and desorption data have been also fitted with different models, and the best fit has been obtained using Avrami's model, as it is capable of accounting for a hybrid CO_2 adsorption mechanism stemming from two different pathways, i.e., physisorption and chemisorption [32].

The experimental campaigns conducted on CB/FM hybrids [28] highlighted that also in CO_2 adsorption under dynamic conditions, the surface area of the carbonaceous support and, more specifically, the distribution of micro- and meso-pores play a relevant role. To go deep into this issue, raw CB has been worked out to expose a different porosity (and surface chemistry features) and explored as a solid support for different typologies of CO_2 active phases [29, 30]. CB surface area was increased by oxidation in hot nitric acid for 24 h from 139 m^2/g (bare CB) up to 323 m^2/g (hereinafter CBox) and the micro-pores/meso-pores ratio modified increasing the volume of micro-pores. CBox was (i) tested as it is; (ii) thermally threated (annealing up to 800°C under inert atmosphere); (iii) grafted with amino groups, CBox-NH_2; (iv) coated with iron oxide particles, CBox/FM (50 wt.%); and (v) impregnated with a triazole-based (trihexyl(tetradecyl)phosphonium triazole, [P66614][Triz]) ionic liquid (IL), CBox/IL. In the latter case, the sample obtained impregnating raw CB with IL (CB/IL) has been also produced for comparison [29, 30]. All the CB-based adsorbents were structurally and morphologically characterized [29]. The materials exhibit a good thermal stability (up to 600°C in the case of oxidized material, up to 200°C in the case of amino-functionalized material and IL-supported adsorbents, and up to 500°C in the case of iron oxide-containing composites) and a good variability in terms of surface areas and pore size distributions [29]. The CO_2 capture performances, evaluated in dynamic conditions in a lab-scale fixed bed micro-reactor operating at CO_2 3 vol.% (inlet flow rate 15 Nl/h) and atmospheric pressure [29], allowed to estimate the material adsorption performances in dependence of the different textural properties of the carbonaceous support.

It was shown (**Figure 4**) that CBox and thermally annealed CBox exhibit lower sorption (8.8 and 11.1 mg_{CO2}/g, respectively), the former comparable to CB [29, 30]. The sample obtained by CBox amine functionalization exhibits high CO_2 uptake despite its low surface area (29.5 m^2/g) indicating a clear chemisorption adsorption mechanism [30].

The comparison among the samples obtained by covering CBox and CB with a CO_2 active phase (IL or FM) clearly indicates that the enhanced CBox micro-porosity greatly limits the accessibility of CO_2 toward the absorbing phase that is clogged into the micro-pores, lowering the number of available binding sites for CO_2 [29]. This effect is remarkable since the CO_2 uptake lowers from 18.3 mg_{CO2}/g to 8.0 mg_{CO2}/g in the case of FM-covered carbons (while the surface area increases from 156.7 in CB/FM to 247.6 m^2/g in CBox/FM) and from 27.3 mg_{CO2}/g to

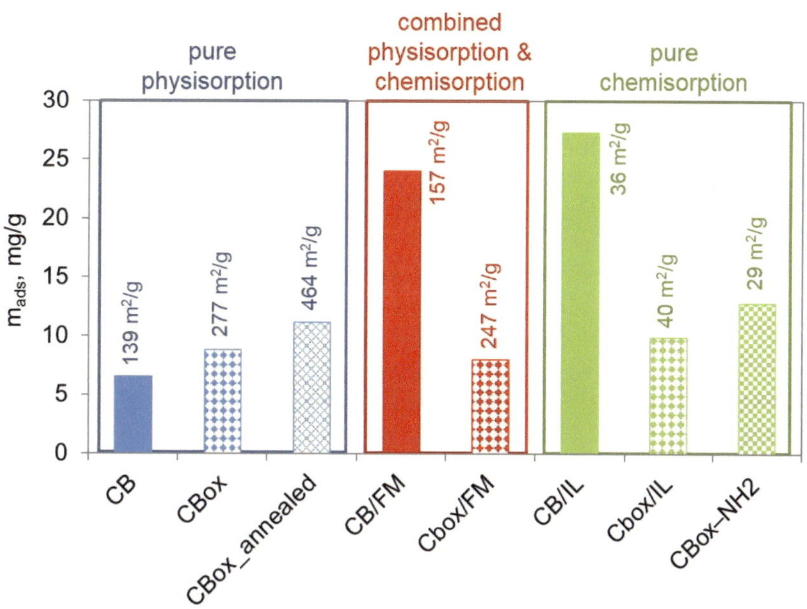

Figure 4.
CO_2 sorption capacities (m_{ads}, mg/g) of the CB-derived hybrid materials. Experimental conditions: 0.35 g of adsorbent, lab-scale fixed bed micro-reactor, 3% CO_2 (inlet flow rate 15 Nl/h) in N_2, room temperature, and pressure.

9.9 mg_{CO2}/g in the case of IL-covered carbons (where the surface area keeps quite constant moving from 36.5 to 40.2 m^2/g in CBox/FM and CBox/FM, respectively) (**Figure 4**) [29].

3. Highly CB destructuration

3.1 Graphene-like (GL) layers and ultrathin (GL) films

The destructuration of the CB microstructure (achieved after 90 h of hot acidic treatment) allows obtaining two classes of water-stable materials placeable into the family of graphene-related materials (GRMs): (i) graphene-like (GL) layers [33, 34] and (ii) small graphenic fragments classifiable as carbon nanodots (CNDs) [35] exhibiting fluorescence in the visible region. The former class (hereinafter GL layers) can be easily obtained by chemical reduction of the product obtained after 90 h of acidic treatment [33, 34]. At the end of the oxidation step, the precursors of GL layers are recovered as a solid after centrifugation, while CNDs are recovered in the surnatant. The GL layer precursors are then converted into GL layers through a chemical reduction in mild conditions [33, 34]. The two-step (oxidation and reduction) approach is performed in aqueous medium and allows scalability and potentially large-amount production at reduced costs.

The morphological characterization of GL layers, performed by atomic force microscopy (AFM) on very diluted GL water suspension (0.1 μg/mL) [36], allowed to image rather individual GL layer units with vertical sizes ranging from about 1 nm or less to a few nanometers; the larger lateral dimensions, a few tens of nanometers, indicate that some aggregations are still occurring and that the primary aggregation process occurs to form preferentially extended thin planes [33].

Unlike rGO, whose basal planes are damaged at a different extent (formation of defects or holes) upon the removal of carbonyl and ether groups, the GL layers are characterized by intact graphenic basal planes. The edges of the GL graphenic layers are mainly decorated with a complex variety of oxygen functional groups (mainly carboxylic and carbonyl groups [33, 34]) giving to the GL layers an exceptional colloidal stability over a wide pH range (3–14) hardly achieved by conventional rGO without using stabilizing agents.

The quantification of the oxygen functional groups by coulometric-potentiometric titration in the pH range of 2.7–7 [34] allows identifying two dominant classes in the carboxylate region (pK 2.0–5.0) with $pKa = 3.40 \pm 0.05$ (number of sites = 900 ± 30 μmol/g) and $pKa = 5.5 \pm 0.1$ (number of sites = 240 ± 30 μmol/g, mainly lactones and carboxylic anhydride groups) [34].

An intriguing aspect of GL is related to its wide chemical versatility that allowed the production of different kinds of hybrid materials. Different materials have been produced and explored for a wide array of applications: metal organic frameworks (MOF)/GL hybrids have been obtained allowing MOF to grow in presence of GL layers [37], biocompatible hybrids have been developed interfacing eumelanin and GL layers (EUGLs) [38, 39], and photocatalytic hybrids have been produced by interfacing TiO_2 and GL layers [40, 41].

Interestingly enough, the GL layers exhibit a peculiar self-assembling behavior when dried on surfaces, arranging in different morphologies (ranging from atomically flat to patterned surface) as a consequence of the suspension pH, surface roughness, and chemistry [34]. Thanks to their peculiar characteristics, GL layers are also particularly suitable to homogenously coat non-planar surfaces, including particles (TiO_2, silica, Al_2O_3) and living cells [36]. When GL layers are contacted with planktonic *S. aureus* living cells [36], they exhibit antimicrobial activity and are also capable to inhibit the formation of *S. aureus* biofilm. It was found that the antimicrobial capabilities were exerted without altering the integrity of the plasma membrane, reasonably by simply coating the cell membrane likely mediated by electrochemical interactions between the GL-derived functional groups and cell wall/plasma membrane proteins. The scavenging activity of GL layers has also been found [40], while bare rGO was beyond the detection limit. Electron paramagnetic resonance (EPR) spectroscopy investigation on GL layers showed a single and asymmetric signal at g-values of ~2.0035 (spin density = 4.3×10^{18} spin/g, signal amplitude $\Delta B = 3.4 \pm 0.2$) typical of carbon-centered radicals and related to the localized spins generated at the edge of the π-electron system [40, 41].

The presence of quite intact basal plane in GL layers guarantees electrical conductivity. I-V characterization showed linear responses implying an Ohmic behavior of GL films. The conductivity of a GL film of around 20 nm height, estimated by the classical four-probe van der Pauw configuration [42], was $2.5 \pm 0.3 \times 10^{-2}$ S/cm [33]. This characteristic was successfully exploited to prepare conductive hybrid materials in water suspension [37–41].

3.2 VOC sensing

The good conductivity exhibited by GL layers combined with the tendency to form differently patterned films was exploited also for the development of prototypal devices for sensoristic applications in volatile organic compound (VOC) detection, paying a particular attention to alcohol detection [43, 44].

The need of low-cost, eco-compatible VOC sensors for the air quality monitoring working at room temperature [45] pushes research activities toward the constant development and testing of new sensing layers. In the manifold of nanomaterials suitable for developing a gas sensor, GRMs attract increasing interest, since they combine excellent detection sensitivity with transduction properties and easily tunable chemo-physical characteristics [45, 46].

The performances of GL films as sensing layers were probed at first using a prototypal chemoresistive device prepared by drop-casting GL solution onto alumina-based transducers and dried at room temperature. The ability to detect alcohol vapors (ethanol and n-butanol) was evaluated at atmospheric pressure, room temperature, and dried atmosphere, in the range of 0–100 ppm. After conditioning for 30 min in dry air (baseline), the sensing device was exposed to analytes for a time ranging between 10 and 30 min and then restored for a time ranging between 20 and 60 min (**Figure 5**).

The GL layer response toward ethanol depends on the reduction strategy adopted (hydrazine hydrate or NMP refluxing [43]) and, expressed as normalized relative variation of conductance, it varies from 3% to 6% for ethanol at 50 ppm and from 0.5% to 1.1% for n-butanol (**Figure 5**) at 50 ppm, where the lower values are referred to GL layers reduced with hydrazine hydrate and the higher responses are referred to GL layers reduced with NMP refluxing [43]. The response of the GL precursors was also reported for comparison (**Figure 5**).

Overall, the lower response to n-butanol than that to ethanol could be ascribed to the different ability of both alcohols to permeate the GL flakes (the analytes have a different steric hindrance), as well as to the different chemical affinity to functional groups on the GL layer surfaces. A combined role of morphology and surface chemistry in determining the response to the analytes was assumed as a possible interpretation of experimental evidences [43].

The device was also tested at increasing concentration steps of ethanol and butanol [43] up to a concentration of 100 ppm (lower panels of **Figure 5**). The tests clearly showed that the sensing response increased according to the increasing

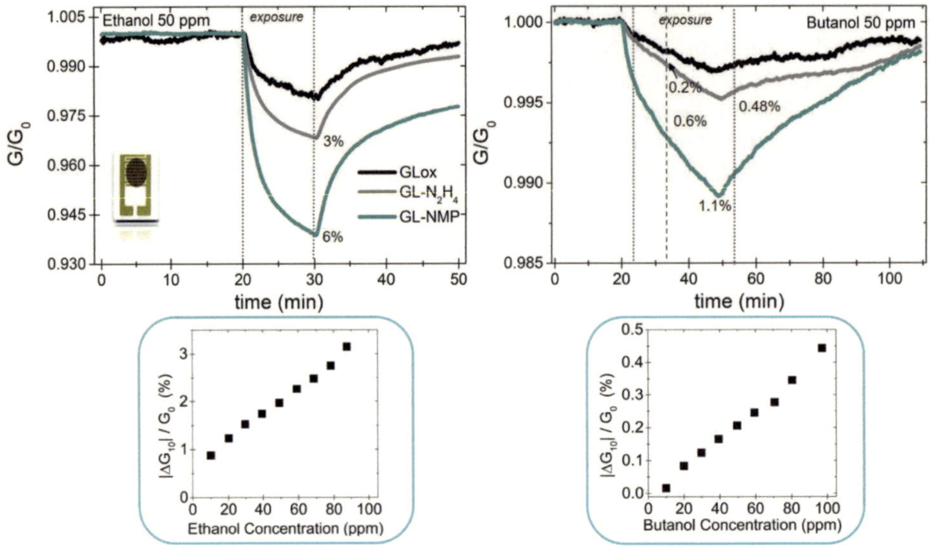

Figure 5.
Responses of the chemoresistive devices based on GLox (black curve), GL-N_2H_4 film (gray curve), and GL-NMP film (blue curve) toward ethanol (upper panel) and n-butanol (lower panel) vapors. The device responses at increasing concentration (up to 100 ppm) of ethanol and butanol are reported in the lower panels. Adapted from [43].

analyte concentration [43] and that the prototypal devices are able to discriminate between different concentrations. The higher sensitivity toward ethanol is confirmed at all the concentrations explored reaching, in the case of GL layers reduced with NMP refluxing, around 1% at 10 ppm of ethanol.

The comparison of the experimental results with those reported by literature showed that GL films are a promising candidate for the detection of low concentrations of ethanol at room temperature [43].

Starting on these encouraging results, preliminary attempts to process GL layers from aqueous suspension in finely controlled paths by means of inkjet printing (IJP) manufacturing have been performed [44]. The IJP technology is a deposition method from liquid phase that well addresses the demand for fine patterning necessary for sensoristic applications [47]. The IJP technology permits an efficient use of nonflexible and flexible eco-friendly substrates (as paper substrates) reducing the amount of waste products [47] by ensuring i) effective control on the deposed ink volume; ii) no-vacuum, no-temperature, and contactless deposition; iii) high reproducibility [47].

Prototypal chemoresistors have been fabricated by printing GL layers onto glossy paper with interdigitated Cr/Au electrodes as a rectangular surface [44]. The printing parameters have been optimized in order to obtain uniform, conductive, and reproducible sensing films (**Figure 6**). In order to assess the reproducibility of the deposition technique, two twin devices were fabricated, and in both devices the sensing GL-based film has been realized by printing multiple overlapped layers.

The sensing properties of the printed GL layer prototypal chemoresistors have been investigated by exposing them to vapors of ethanol under the same conditions reported above (**Figure 6**). The response toward ethanol (50 ppm, in dry N_2) in terms of conductance variation was 1.3%. Intra-substrate reproducibility in terms of surface morphology (macroscopic and microscopic distribution of the printed nanomaterial) and electrical responses (base resistance and conductance variation) was also demonstrated [44].

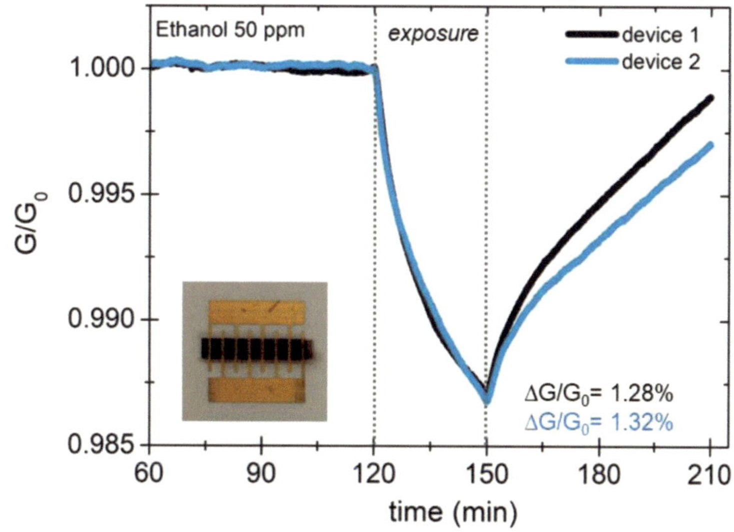

Figure 6.
Responses of the twin GL-N_2H_4-based chemoresistive devices produced by IJP toward ethanol vapors. Adapted from [44].

The possibility to use different substrates (alumina, silicon dioxide) is currently under study with the aim of evaluating the more suitable substrate for GL-based chemoresistor production.

4. Conclusions

This chapter reports the very recent experimentation aimed to valorize the potentialities of carbon black for practical and uncommon applications. The proposed approaches are purposively designed to be simple, cost-effective, versatile, suitable for synthesis scale-up, and, possibly, sustainable (quite mild synthesis conditions, where water is mostly used as solvent). The approaches have been developed taking into account the relationships between CB-derived carbons nanostructuration (i.e., micro-/meso-pore ratio, specific surface chemistry, etc.) and macroscopic properties.

It was shown that CB can be modified on the surface (preserving the microstructure) and fully destructurated until separating the graphenic building blocks embedded in. The preserved conductive properties allow the exploitation of the graphenic layers (graphene-like layers) for VOC sensing application, thus avoiding the use of metal oxides for this specific application in view of more sustainable sensors. Applications in CO_2 capture and storage (CCS) and water remediation (heavy metal capture and rare-earth element recovery) of CB modified at the surface resulted encouraging and mature for practical applications.

Acknowledgements

This paper is based on the collaboration with Roberto Di Capua, Fulvia Villani, Tiziana Polichetti, Francesco di Natale, Paola Ammendola, Federica Raganati, and Luciana Lisi since many years.

Conflict of interest

The authors declare no conflict of interest.

Author details

Michela Alfè* and Valentina Gargiulo
Institute for Research on Combustion (IRC-CNR), Naples, Italy

*Address all correspondence to: alfe@irc.cnr.it

References

[1] Long CM, Nascarella MA, Valberg PA. Carbon black vs. black carbon and other airborne materials containing elemental carbon: Physical and chemical distinctions. Environmental Pollution. 2013;**181**:271-286. DOI: 10.1016/j.envpol.2013.06.009

[2] Fan Y, Fowler GD, Zhao M. The past, present and future of carbon black as a rubber reinforcing filler: A review. Journal of Cleaner Production. 2020;**247**:119115. DOI: 10.1016/j.jclepro.2019.119115

[3] Watson AY, Valberg PA. Carbon black and soot: Two different substances. AIHA Journal. 2001;**62**(2):218-228. DOI: 10.1080/15298660108984625

[4] Arnal C, Alfè M, Gargiulo V, Ciajolo A, Alzueta MU, Millera A, et al. Characterization of soot. In: Battin-Leclerc F, Simmie JM, Blurock E, editors. Cleaner Combustion-Developing Detailed Chemical Kinetic Models. Springer Nature Switzerland; 2013. pp. 333-362

[5] Boehm HP. Some aspects of the surface chemistry of carbon blacks and other carbons. Carbon. 1994;**32**(5):759-769. DOI: 10.1016/0008-6223(94)90031-0

[6] Mannfredi C, Mozzillo R, Volino S, Trifuoggi M, Giarra A, Gargiulo V, et al. On the modeling of heavy metals and rare earth elements adsorption on colloidal carbon-based nanoparticles. Applied Surface Science. 2020;**505**:144264. DOI: 10.1016/j.apsusc.2019.144264

[7] Lounasvuori MM, Kelly D, Foord JS. Carbon black as low-cost alternative for electrochemical sensing of phenolic compounds. Carbon. 2018;**129**:252-257. DOI: 10.1016/j.carbon.2017.12.020

[8] Wu C-S, Chang T-W, Teng H, Lee Y-L. High performance carbon black counter electrodes for dye-sensitized solar cells. Energy. 2016;**115**:513-518. DOI: 10.1016/j.energy.2016.09.052

[9] Liu D, Zhao X, Song L, Zhang S. Synthesis, performance and action mechanism of carbon black/Ag_3PO_4 photocatalysts. Ceramics International. 2018;**44**:13712-13719. DOI: 10.1016/j.ceramint.2018.04.212

[10] Kokhanovskaya OA, Razdyakonova GI, Likholobov VA. New applications of carbon black. An aerogel-like composite material with heat insulating properties. Procedia Engineer. 2016;**152**:540-544. DOI: 10.1016/j.proeng.2016.07.652

[11] Pahalagedara LR, Siriwardane IW, Tissera ND, Wijesena RN, Nalin de Silva KM. Carbon black functionalized stretchable conductive fabrics for wearable heating applications. RSC Advances. 2017;**7**:19174-19180. DOI: 10.1039/C7RA02184D

[12] Chen B, Li B, Gao Y, Ling T-C, Lu Z, Li Z. Investigation on electrically conductive aggregates produced by incorporating carbon fiber and carbon black. Construction and Building Materials. 2017;**144**:106-114. DOI: 10.1016/j.conbuildmat.2017.03.168

[13] Monteiro AO, Loredo A, Costa PMFJ, Oeser M, Cachim PB. A pressure-sensitive carbon black cement composite for traffic monitoring. Construction and Building Materials. 2017;**154**:1079-1086. DOI: 10.1016/j.conbuildmat.2017.08.053

[14] Santini E, Ravera F, Ferrari M, Alfè M, Ciajolo A, Liggieri L. Interfacial properties of carbon particulate-laden liquid interfaces and stability of related foams and emulsions. Colloid Surface A. 2010;**365**(1-3):189-198. DOI: 10.1016/j.colsurfa.2010.01.041

[15] Gargiulo V, Alfè M, Lisi L, Manfredi C, Volino S, Di Natale F. Colloidal carbon-based nanoparticles as heavy metal adsorbent in aqueous solution: Cadmium removal as a case study. Water, Air, & Soil Pollution. 2017;**228**(5):192-205. DOI: 10.1007/s11270-017-3378-5

[16] Alfe M, Gargiulo V, Di Capua R. An old but lively nanomaterial: Exploiting carbon black for the synthesis of advanced materials. Eurasian Chemico-Technological Journal. 2019;**21**(3):203-213. DOI: 10.18321/ectj861

[17] Di Natale F, Gargiulo V, Alfè M. Adsorption of heavy metals on silica-supported hydrophilic carbonaceous nanoparticles (SHNPs). Accepted on Journal of Hazardous Materials. 2020. DOI: 10.1016/j.jhazmat.2020.122374

[18] Decreto Legislativo 3 aprile 2006, n. 152. Available from: http://www.camera.it/parlam/leggi/deleghe/06152dl.htm

[19] Alfarra A, Frackowiak E, Béguin F. The HSAB concept as a means to interpret the adsorption of metal ions onto activated carbons. Applied Surface Science. 2004;**228**:84-92. DOI: 10.1016/j.apsusc.2003.12.033

[20] Leung DYC, Caramanna G, Maroto-Valer MM. An overview of current status of carbon dioxide capture and storage technologies. Renewable and Sustainable Energy Reviews. 2014;**39**:426-443. DOI: 10.1016/j.rser.2014.07.093

[21] Samanta A, Zhao A, Shimizu GKH, Sarkar P, Gupta R. Post-combustion CO_2 capture using solid sorbents: A review. Industrial and Engineering Chemistry Research. 2012;**51**:1438-1463. DOI: 10.1021/ie200686q

[22] D'Alessandro DM, Smit B, Long JR. Carbon dioxide capture: Prospects for new materials. Angewandte Chemie International Edition. 2012;**49**:6058-6082. DOI: 10.1002/anie.201000431

[23] Nugent P, Belmabkhout Y, Burd SD, Cairns AJ, Luebke R, Forrest RK, et al. Porous materials with optimal adsorption thermodynamics and kinetics for CO_2 separation. Nature. 2013;**495**:80-84. DOI: 10.1038/nature11893

[24] Oschatz M, Antonietti M. A search for selectivity to enable CO_2 capture with porous adsorbents. Energy & Environmental Science. 2018;**11**:57-70. DOI: 10.1039/C7EE02110K

[25] Baltrusaitis J, Jensen JH, Grassian VH. FTIR spectroscopy combined with isotope labeling and quantum chemical calculations to investigate adsorbed bicarbonate formation following reaction of carbon dioxide with surface hydroxyl groups on Fe_2O_3 and Al_2O_3. The Journal of Physical Chemistry. B. 2006;**110**:12005-12016. DOI: 10.1021/jp057437j

[26] Mishra AK, Ramaprabhu S. Nano magnetite decorated multiwalled carbon nanotubes: A robust nanomaterial for enhanced carbon dioxide adsorption. Energy & Environmental Science. 2011;**4**:889-895. DOI: 10.1039/C0EE00076K

[27] Mishra AK, Ramaprabhu S. Magnetite decorated graphite nanoplatelets as cost effective CO_2 adsorbent. Journal of Materials Chemistry. 2011;**21**:7467-7471. DOI: 10.1039/C1JM10996K

[28] Alfè M, Ammendola P, Gargiulo V, Raganati F, Chirone R. Magnetite loaded carbon fine particles as low-cost CO_2 adsorbent in a sound assisted fluidized bed. Proceedings of the Combustion Institute. 2015;**35**:2801-2809. DOI: 10.1016/j.proci.2014.06.037

[29] Gargiulo V, Alfè M, Ammendola P, Raganati F, Chirone R. CO_2 sorption

on surface-modified carbonaceous support: Probing the influence of the carbon black microporosity and surface polarity. Applied Surface Science. 2016;**360**:329-337. DOI: 10.1016/j.apsusc.2015.11.026

[30] Gargiulo V, Alfè M, Raganati F, Zhumagaliyeva A, Doszhanov Y, Ammendola P, et al. CO_2 adsorption under dynamic conditions: An overview on rice husk-derived sorbents and other materials. Combustion Science and Technology. 2018;**191**(9):1484-1498. DOI: 10.1080/00102202.2018.1546697

[31] Raganati F, Alfe M, Gargiulo V, Chirone R, Ammendola P. Isotherms and thermodynamics of CO_2 adsorption on a novel carbon-magnetite composite sorbent. Chemical Engineering Research and Design. 2018;**134**:540-552. DOI: 10.1016/j.cherd.2018.04.037

[32] Raganati F, Alfe M, Gargiulo V, Chirone R, Ammendola P. Kinetic study and breakthrough analysis of the hybrid physical/chemical CO_2 adsorption/desorption behavior of a magnetite-based sorbent. Chemical Engineering Journal. 2019;**372**:526-535. DOI: 10.1016/j.cej.2019.04.165

[33] Alfè M, Gargiulo V, Di Capua R, Chiarella F, Rouzaud J-N, Vergara A, et al. Wet chemical method for making graphene-like films from carbon black. ACS Applied Materials & Interfaces. 2012;**4**:4491-4498. DOI: 10.1021/am301197q

[34] Alfè M, Gargiulo V, Di Capua R. Tuning the surface morphology of self-assembled graphene-like thin films through pH variation. Applied Surface Science. 2015;**353**:628-635. DOI: 10.1016/j.apsusc.2015.06.117

[35] Dong Y, Chen C, Zheng X, Gao L, Cui Z, Yang H, et al. One-step and high yield simultaneous preparation of single- and multi-layer graphene quantum dots from CX-72 carbon black. Journal of Materials Chemistry. 2012;**22**:8764-8766. DOI: 10.1039/c2jm30658a

[36] Olivi M, Alfè M, Gargiulo V, Valle F, Mura F, Di Giosia M, et al. Antimicrobial properties of graphene-like nanoparticles: Coating effect on *Staphylococcus aureus*. Journal of Nanoparticle Research. 2016;**18**(12):358. DOI: 10.1007/s11051-016-3673-x

[37] Alfè M, Gargiulo V, Lisi L, Di Capua R. Synthesis and characterization of conductive copper-based metal-organic framework/graphene-like composites. Materials Chemistry and Physics. 2014;**147**:744-750. DOI: 10.1016/j.matchemphys.2014.06.015

[38] Gargiulo V, Alfè M, Di Capua R, Togna AR, Cammisotto V, Fiorito S, et al. Supplementing π-systems: Eumelanin and graphene-like integration towards highly conductive materials for the mammalian cell culture bio-interface. Journal of Materials Chemistry B. 2015;**3**:5070-5079. DOI: 10.1039/C5TB00343A

[39] Di Capua R, Gargiulo V, Alfè M, De Luca GM, Skála T, Mali G, et al. Eumelanin graphene-like integration: The impact on physical properties and electrical conductivity. Frontiers in Chemistry. 2019;**7**:121. DOI: 10.3389/fchem.2019.00121

[40] Lettieri S, Gargiulo V, Pallotti DK, Vitiello G, Maddalena P, Alfè M, et al. Evidencing opposite charge-transfer processes at TiO_2/graphene-related materials interface through a combined EPR, photoluminescence and photocatalysis assessment. Catalysis Today. 2018;**315**:19-30. DOI: 10.1016/j.cattod.2018.01.022

[41] Alfè M, Spasiano D, Gargiulo V, Vitiello G, Di Capua R, Marotta R. TiO_2/graphene-like photocatalysts for selective oxidation of 3-pyridine-methanol to

vitamin B3 under UV/solar simulated radiation in aqueous solution at room conditions: The effect of morphology on catalyst performances. Applied Catalysis A: General. 2014;**487**:91-99. DOI: 10.1016/j.apcata.2014.09.002

[42] Van Der Pauw LJ. A method of measuring specific resistivity and Hall effect of discs of arbitrary shape. Philips Research Reports. 1958;**13**:1-9

[43] Gargiulo V, Alfano B, Di Capua R, Alfé M, Vorokhta M, Polichetti T, et al. Graphene-like layers as promising chemoresistive sensing material for detection of alcohols at low concentration. Journal of Applied Physics. 2018;**123**(2):024503. DOI: 10.1063/1.5000914

[44] Villani F, Loffredo F, Alfano B, Miglietta ML, Verdoliva L, Alfè M, et al. Graphene-like based-chemoresistors inkjet-printed onto paper substrate. In: Leone A, Forleo A, Francioso L, Capone S, Siciliano P, Di Natale C, editors. Sensors and Microsystems: Proceedings of the 19th AISEM 2017 National Conference (Lecture Notes in Electrical Engineering). Springer Nature Switzerland; 2019

[45] Baptista FR, Belhout SA, Giordani S, Quinn SJ. Recent developments in carbon nanomaterial sensors. Chemical Society Reviews. 2015;**44**:4433-4453. DOI: 10.1039/C4CS00379A

[46] Llobet E. Gas sensors using carbon nanomaterials: A review. Sensors and Actuators, B: Chemical. 2013;**179**:32

[47] Singh M, Haverinen HM, Dhagat P, Jabbour GE. Inkjet printing—Process and its applications. Advanced Materials. 2010;**22**:673-685. DOI: 10.1002/adma.200901141

Chapter 5

An Overview of Carbon-Based Materials for the Removal of Pharmaceutical Active Compounds

Mazen K. Nazal

Abstract

Carbon-based materials, namely activated carbon, carbon nanotube and graphene, are considered as one of the most effective adsorbents for pollutant removal and wastewater treatment. Due to their high surface area and distinct chemical and physical properties of the carbon-based materials, particularly activated carbon and carbon nanotube are rapidly emerging as one of the most effective adsorbents for wastewater treatment. Various studies have reported the applications of activated carbon, carbon nanotubes and graphene as promising adsorbents for removing organic and inorganic pollutants. In this chapter, an introduction about the activated carbon, carbon nanotubes and graphene and their production, prosperities and usage for the removal of pharmaceutical active materials from aqueous media are highlighted and summarized. Challenges and future opportunities for application of these carbon-based materials as adsorbents in wastewater treatment are also addressed in this chapter.

Keywords: adsorption, aquatic environment, pollutants, activated carbon, carbon nanotube

1. Introduction

In recent years, there is great concern about the occurrence and the impact of the pharmaceutical active compounds in water, in addition, development of efficient and cost-effective technologies for the removal of these compounds and treatment of industrial effluent, surface water and ground water. Pharmaceutical active compounds are natural or synthetic chemicals that can be found in over-the-counter therapeutic drugs and veterinary drugs. They induce pharmacological effect and give significant benefits to human beings. A continuous release of these chemical compounds into aquatic environment has been increased due to the increase of general use of pharmaceutical compounds in human and veterinary medicines. **Figure 1** illustrates the routes of releasing the pharmaceutical compounds into water. These routes include wastewater effluents, human and animal excreta, sewage sludge, medical and industrial waste and land fill leaching [1].

Depending on the biodegradability and hydrophobicity of these pharmaceutical active compounds, they are naturally reduced by dilution, degradation and

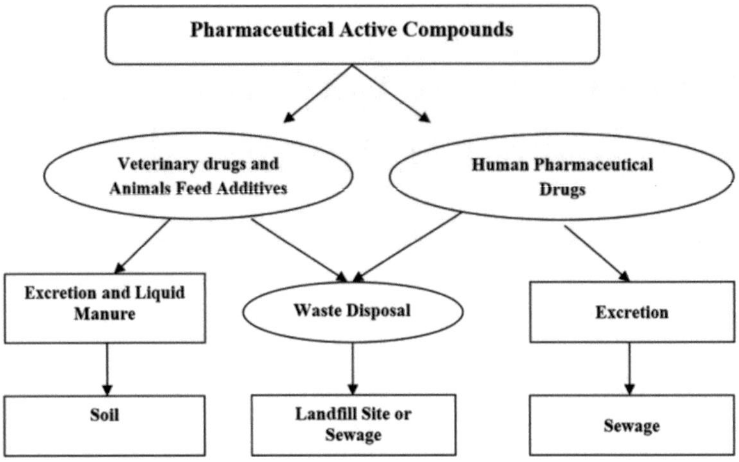

Figure 1.
Routes of releasing the pharmaceutical compounds into the environment.

Pharmaceutical active compounds	Maximum detected concentration (ng/L)	Aquatic environment type
Bleomycin	19 (United Kingdom)	Sewage
Clotrimazole	34 (United Kingdom)	Stream or river water
Diclofenac	1200 (Germany) 41 (France) 40 (Finland) 64 (Austria)	Surface water
Carbamazepine	110 (Germany) 800 (France) 370 (Finland) 64 (Austria)	Surface water
Iopromide	910 (Germany) 17 (France) 211 (Austria)	Surface water
Roxithromycin	560 (Germany) 37 (France)	Surface water
Ibuprofen	530 (Germany) 120 (France) 65 (Finland)	Surface water
Erythromycin	80 (United Kingdom)	River water
Fluoxetine	290 (United Kingdom)	Sewage
Mefenamic acid	1440 (United Kingdom)	Sewage
Paracetamol	< 20 (United Kingdom)	Sewage
Propranolol	215 (United Kingdom)	River water
Tamoxifen	42 (United Kingdom)	Sewage
Tetracycline	1000 (United Kingdom)	River water
Trimethoprim	1288 (United Kingdom)	Sewage

Table 1.
The measured concentration of some pharmaceutical active compounds in some of the aquatic environment in European countries.

adsorption in the environment. Thus, these compounds in water exist in a trace concentration level [2].

Some of the pharmaceutical active compounds used for birth control, heart medication and painkilling were detected in wastewater in the United State of America (USA) since more than 40 years ago [3–5]. Literature shows that the pharmaceutical active compounds enter the surface water through different sources such as excretion, bathing, effluent discharging, improper disposal of these compounds and veterinary facilities [1, 6–8]. In addition, a study conducted in the United Kingdom by Drinking Water Inspectorate reported that many classes of pharmaceutical active compounds are present in wastewater influent [9]. **Table 1** represents several pharmaceutical active compounds that were detected in the aquatic environment of United Kingdom (UK) and other European countries [10, 11].

There is no international standard method for drinking water sampling and method of analysis for pharmaceutical active compounds. In addition, a few systematic monitoring studies on measuring the pharmaceutical active compounds in surface water, drinking water and ground water were conducted. Therefore, limited data are available on their occurrence in these aquatic environments to be used in assessing the potential health risk due to the exposure to a trace concentration level of pharmaceutical compounds. However, literature showed that the surface water and ground water sources affected by wastewater discharges have pharmaceutical active compound concentrations less than 100 ng/L, while these compounds were found in the drinking water with a concentration less than 50 ng/L [2].

2. Treatment technologies for pharmaceutical compounds' removal from water

The presence of these compounds at trace concentration levels (nanogram to sub microgram per liter) in the aquatic environment has raised a question concerning the efficiency of wastewater treatment techniques in removing of the pharmaceutical active compounds. Many removal techniques such as chlorination, photocatalysis, adsorption, biodegradation and advanced oxidation or ozonation have been investigated for the removal of pharmaceutical active compounds from the aquatic environment [12–25]. Some of these techniques have different disadvantages such as their high cost, high energy consumption and formation of toxic by-products. Adsorption technique has many advantages over these techniques such as it works at mild operation conditions, requires low energy and is efficient and cost-effective. Therefore, it is a promising technique for the removal of pharmaceutical active compounds.

2.1 Adsorption technique

The removal of pharmaceutical active compounds from water by adsorption is considered as one of the easiest and safest techniques since it is easy to design and operate and this technique does not produce any toxic wastes as a by-product and is capable of removing most forms of organic material. The adsorption process includes the accumulation of pharmaceutical compounds on the adsorbent's surface. Hence, the selection of adsorbent must be precious. The adsorbent must have a capability to accumulate the pollutant from water with high surface area and high hydrophobicity. The efficiency of this technique is mainly depending on the functional group composition, surface area, pore size and the ash content. It also depends on the chemical parameters like temperature, polarity, pH, concentration of the adsorbate and the availability of other competing solutes. The adsorption process also depends on the mobility of the adsorbate molecules toward the external boundary layer of the adsorbent, active surface sites and surface pore size.

Pharmaceutical active compound	Carbon-based adsorbent	Adsorption capacity (mg/g)	Reference
Clofibric acid	Mesoporous silica SBA-15	70	[70]
Ofloxacin	Nonporous SiO_2	2.1	[71]
Tetracycline	Mesoporous silica	44.4	[27]
Cephalexin	Amberlite XAD-16 polymer	116	[36]
Nalidixic acid	Polystyrene-divinylbenzene, X16	800	[31]
Penicillin	Polymer Amberlite XAD-16	1401	[34]
Amoxicillin	Bentonite clay	53.9	[38]
Flurbiprofen	Organophilic montmorillonite clay	240	[39]
Tetracycline	Na-kaolinite	29	[40]
	Kaolinite	3.8	[72]
	Rectorite clay	40	[46]
Tetracycline	NaOH-activated carbon produced from macadamia nut shells	455.33	[48]
	H_3PO_4-activated carbon produced from apricot nut shells	308.3	[49]
	Activated carbons produced by KOH activation of tyre pyrolysis char	312	[50]
	Commercial activated carbon	471	[51]
Sulfamethoxazole	AC	185	[53]
Metronidazole	AC	93.21	[53]
	CAC	328	[52]
Amoxicillin	AC	221.8	[73]
Dimetridazole	CAC	186	[52]
Ronidazole	CAC	394	[52]
Tinidazole	CAC	385	[52]
Penicillin G	AC	315	[56]
Oxytetracycline	MWNT10	190.2	[54]
Tetracycline	MWNTs	148	[58]
	SWNTs	370	
Tylosin	K-MWNTs	270	[58]
	K-SWNTs	466	
Carbamazepine	MWNT100	41.4	[58]
Cephalexin	Cellulose oxide	79	[59]
Fluoroquinolone	Goethite	49.6	[61]
Ciprofloxacin	Hydrous oxides of Al (HAO)	13.6	[64]

Table 2.
Different adsorbents and their adsorption capacities for removal of pharmaceutical active compounds.

Many researchers have studied the adsorption of pharmaceutical active compounds from wastewater using different types of adsorbents. Several materials as an adsorbent have been reported in the literature and listed in **Table 2** and were tested and investigated for the pharmaceutical active compounds' removal from aquatic environment, such as silica-based adsorbents [26–30], polymeric materials [31–37], clay [38–47], carbonaceous materials [48–58] and other materials [59–71]. The next sections focus on carbonaceous materials as adsorbents, namely activated carbon and carbon nanotubes.

2.1.1 Activated carbon

Activated carbon is a pure carbon graphite form with amorphous and highly porous structure. It contains different range of pore sizes starting from cracks to slits of molecular dimensions [73]. The first produced commercially activated carbon was in early nineteenth century from wood as a raw material. It has been used for water odor and taste control in 1930 [74]. Nowadays, activated carbon is produced from a wide range of raw organic materials and sources, such as sugar, shells, refinery coke, rice hulls and different types of wood. The main features of activated carbon that make it good as an adsorbent in the adsorption process are the following: (i) its high surface area, (ii) its porosity and (iii) its surface reactivity.

2.1.1.1 Classifications of activated carbons

Activated carbon can be classified based on its activation process or its properties. Based on the activation process, the following are the main two categories based on the activation process:

- Physically or thermally activated carbon: the activation process involves carbonization of organic raw materials at temperature ranging from 500°C to 600°C [75].
- Chemically activated carbon: the activation process involves addition of some inorganic salts such as metallic chloride to activate the surface of carbon [76].

Mattson et al. [77] suggested another classification, which categorizes activated carbon to acidic or basic activated carbon:

- Carbon activated at low temperature range from 200°C to 400°C: this develops an acidic surface that lowers the pH value of the solution. This activated carbon exhibits negative zeta potential and usually adsorbs basic and hydrophilic compounds.
- Carbon activated at a high temperature range from 800°C to 1000°C: this develops basic surface that increases the pH value of the solution. Therefore, this type of activated carbon has a positive zeta potential and is usually used for adsorbing acidic organic compounds.

Commercially, activated carbon can be classified as three main types [78], and they are the following:

- Powdered activated carbon (PAC): it has fine granules or powder with particle size less than 1.0 mm and average diameters ranging between 0.15 and 0.25 mm.

- Granular activated carbon (GAC): it combines powdered activated carbon with a binder and forms cylindrical shape activated carbon particles with diameters from 0.8 to 130 mm. The main application for this form is for gas purification.

- Impregnated activated carbon (IAC): it is impregnated with different inorganic ions.

- Polymeric coated activated carbon, which is used in medical field applications.

2.1.1.2 Physicochemical properties of activated carbon

The properties of activated carbon are influenced by the used raw materials and activation method in its preparation process. The porous graphite and graphene sheets that form the activated carbon are connected together and have π-orbitals in the benzene rings, which enable several modifications to be carried out on activated carbon. For example, cooling the activated carbon in the presence of oxygen can produce activated carbon rich with oxides and acidic functional groups, as a result, alter the positive zeta potential of basic activated carbon to negative to be used for different applications. In addition, the surface chemistry, pore structure (volume and diameter) and surface area of activated carbon depend significantly on the employed temperature in the preparation process [75, 79].

2.1.1.3 Activated carbon production

A wide range of raw materials can be used as a starting material for producing activated carbon as stated in Section 2.1.1. The following activation methods are used in activated carbon production:

- Thermal activation: this physical process may involve two main steps: the first one to eliminate the volatile matters in the raw materials by carbonizing them thermally at a temperature ranging from 500°C to 600°C and in the second step the porosity and surface are improved by the gasification process. In the gasification process, a carbon dioxide CO_2, methane or steam as an oxidizing gas is used at a high temperature of 800–1000°C [75].

- Chemical activation: in this process, inorganic salts such as metallic chloride are added before the carbonization step to improve the micro-porosity as well as the surface area of the activated carbon [76].

2.1.1.4 Activated carbon for removal of pharmaceutical active compounds

Activated carbon (AC) is widely used in adsorption processes as filtration and purification materials. For instance, in water treatment, activated carbon is used to control taste and odor and to adsorb undesired suspended metals and pollutants [74]. Due to the high surface area and commercial availability of AC, it was studied for removal of different pharmaceutical active compounds. **Table 3** summarizes some of these pharmaceuticals. For example, different types of activated carbon were used for removal of tetracycline (antibiotic drug) from aqueous media. Martins et al. [48] prepared activated carbon from macadamia shells as precursors, the yield was 19.79% and the prepared activated carbon's surface area was 1524 m^2/g. They used it for the tetracycline removal and it had 455.33 mg/g adsorption capacity. Muthanna et al. [80] reported that the activated carbon was used for removal of three pharmaceutical active compounds (i.e., tetracycline, penicillins and quinolones) and the used activated carbon has 1340.8 mg/g adsorption capacity for tetracycline. Chen et al. [81] studied the effect of the adsorption parameters (i.e., pH, contact time, initial concentration and temperature) on the removal of tetracycline from aqueous solution using rice husk ash (RHA). They found the adsorption capacity increased from 1.51 to 3.41 mg/g when the initial tetracycline concentration in the solution increased from 5 to 20 mg/l. Another study showed that activated carbon prepared via a chemical activation of apricot shells using phosphoric acid heated in air at 100 °C for 24 hours has 307.6 m^2/g surface area and 308.3 mg/g adsorption capacity [49]. In 2016, an activated carbon (TPC-AC) was prepared from tires waste by their pyrolysis and then activated using potassium hydroxide [50]. The prepared adsorbent was tested for tetracycline removal and it has been found that the adsorption process was spontaneous and has adsorption capacity (312 mg/g) higher than the commercial activated carbon. Carl et al. [51] reported that the adsorption capacity of the commercial activated carbon for tetracycline is directly related to the density of π electrons in the graphene layers on activated carbon and the aromatic ring in the tetracycline.

2.1.2 Carbon nanotube

Single and multiwall carbon nanotube (CNT) materials are graphene sheets rolled-up tubular individually or more than one inside each other. CNTs were discovered by Sumio Ijima in 1991 at NEC Laboratory in Japan using the Arc discharge production method and then characterized using a transmission electron microscope [82]. CNTs have two different structures based on the rolling direction of graphene sheets: (i) armchair nanotube and (ii) zigzag nanotube structure [83] as shown in **Figure 2**.

The cylindrical shape of CNT nanostructure can have a length to diameter ration up to 132,000,000:1, which is significantly higher than any other materials [83]. This property was explained by the sp2 hybridization in the carbon atoms that CNTs are composed of in addition to the natural alignment of CNT into ropes attracted together by Van der Waals interaction [84].

2.1.2.1 Physical properties and chemical reactivity of carbon nanotubes

CNTs form bundles of a highly complex network [85]. They have electrical conductivity that depends on the arrangement of the hexagonal rings along the tubular surface. Due to their extraordinary properties, such as large geometric aspect ratio, nanocavities and electrical conductivity, CNTs are considered as attractive candidates in many nanotechnological applications, including the removal of pharmaceutical compounds in water treatment processes. One of the

Physical property	Material name				
	MWCNTs	SWCNTs	Wood	Steel	Epoxy
Density (g/cm^3)	2.6	2.6	0.6	7.8	1.25
Tensile strength (Gpa)	150	150	0.008	0.4	0.005
Young's modulus (Gpa)	1200	1054	0.6	208	3.5

Table 3.
Comparison between CNTs and other materials.

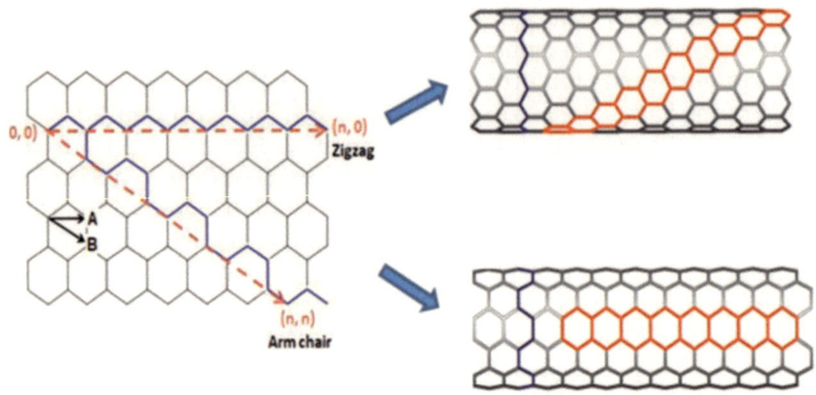

Figure 2.
Armchair and zigzag structural forms of CNTs

main drawbacks of carbon nanotubes is that they do not have good suspension properties in aqueous and organic solvents that in turn has made CNTs' use in industry limited [86]. This disadvantage can be overridden by modifying CNTs chemically with some hydrophilic functional groups that in turn increase CNTs' suspension in water.

The main distinct properties of the carbon nanotubes are categorized into the following:

- Mechanical properties: due to the covalent sp^2 bonds formed between the individual carbon atoms, CNTs have high strength and stiffness. According to the reported results, CNTs have elasticity higher than steel by 10–100 times with an elastic modulus 1Tpa [87]. A comparison between some materials, which have good mechanical properties, with CNTs is shown in **Table 3**.

- Thermal conductivity: CNTs have thermal conductivity ranging from 2800 up to 6000 W/m K [88].

- Electrical properties: CNT carbon-based material exhibits extraordinary electrical properties and it can be conducting or semiconducting material. The conductive CNTs are found to carry electrical current thousand times higher than copper material [89].

- Chemical reactivity: CNTs can chemically be modified to make them highly soluble in aqueous and organic solutions as well as more efficient for certain applications. Their reactivity is related to the mismatching of π-orbitals, which are caused by the curvatures in CNTs' structure. In general, smaller nanotube diameters result in increasing their reactivity. Moreover, the reported results showed that chemical modification of sidewalls or end caps of CNTs are also possible [90].

Based on the CNTs' properties that have been discussed above, CNT materials and their modified structures are promising for different applications such as water treatment, environmental protection and pharmaceutical active compound removal, material science, medicinal chemistry and others.

2.1.2.2 Carbon nanotube production

CNTs are produced using different techniques, and the most common and widely used techniques are:

- **Arc discharge technique.** Arc discharge technique is the most common and simplest technique for CNT production. As mentioned earlier, CNTs were firstly discovered using this technique. In arc discharge technique, CNTs are produced at low pressure of helium inert gas or any other neutral gas [91]. They are produced through arc vaporization of two separated carbon rods in an enclosed system filled with inert gas [92]. One of the major disadvantages of CNT production using this technique is that the produced CNTs are not pure containing some of the catalytic metals; therefore, they require purification to remove these metals and get clean CNTs.

- **Laser ablation technique**. In 1995, carbon nanotubes were synthesized using a laser beam to vaporize graphite at 1200°C [93]. The pulsed and continuous laser methods are the main two types of laser ablations. Much higher light intensity (100 kW/cm^2) is used in the pulsed laser, compared to 12 kW/cm^2 in case of the continuous laser type, which is the main difference between these two laser ablation technique types. In the laser ablation method, CNTs are produced and collected on a cooler surface in the reactor system as the vaporized carbon is condensed. In this technique, SWCNTs can be produced from graphite electrodes by adding metal-based catalysts such as Co, Fe and Ni to the system. However, MWCNTs are the main product when a pure graphite electrode is used [94].

- **Chemical vapor deposition (CVD) technique**. Chemical vapor deposition technique is a simple process and it is believed to be the easiest technique for industrial production of CNTs. In this method, the desired CNT type and quality can be produced by controlling the system production parameters such as temperature, type of catalyst and type of carbon source gases. CVD technique consists of two main steps (catalyst preparation step and then CNT synthesis). In general, to produce CNTs, methane and carbon monoxide gases are dissociated into reactive carbon atoms using an energy source, and then these reactive atoms diffuse over a substrate that is coated by transition metals as a catalyst and heated at a temperature range from 500 to 1000°C [95].
  A comparison between the previously discussed methods for CNT production is summarized in **Table 4**.

2.1.2.3 Carbon nanotubes for pharmaceutical active compounds' removal

Carbon nanotubes with their excellent properties show considerable adsorption capability for removal of pharmaceutical active compounds. A study in 2009 found that the single wall carbon nanotubes (SWNTs) are more efficient for removal of tetracycline from aqueous solutions than multiwall carbon nanotubes (MWNTs), graphite and activated carbon [58]. This finding was explained through the molecular sieving effect, whereas the tetracycline is bulky molecules failed to seep through inner pores, which indicates the important role of molecules' size and their accessibility into pores in the adsorbent materials. In 2016, Yu et al. [96] studied the adsorption performance of MWNTs for removal of ciprofloxacin and found the

Method	Yield (%)	SENT	MINT	Concerns
Chemical vapor deposition	20–100%	Long tubes with diameters 0.6–4 nm	Long tubes with diameters 10–240 nm	Nets are usually mints and often riddled with defects
Arc discharge	30–90%	Short tubes with diameters 0.6–1.4 nm	Short tubes with inner diameter 1–3 nm	Short tubes with random sizes and directions and required purification
Laser ablation (Vaporization)	Up to 70%	Long bundles of tubes with diameters 1–2 nm	Not suitable and too expensive	Costly and required high power

Table 4.
Comparison between the three methods in terms of CNT production efficiency, type of CNTs produced, and the current drawback of each technology.

maximum adsorption capacity is 20 mg/g, which was obtained at pH 4 and 240 min that was attributed by the π-π interaction rather than hydrogen bonding and interaction with oxygenated functional groups on MWNTs. Another study by Yu et al. [97] showed that the maximum adsorption capacity of MWNTs for tetracycline was 269.54 mg/g, which achieved at 25°C and pH 5 within 80 min.

In order to improve the performance and adsorption capacity of CNTs, different types of modifications can be performed such as graphitization, hydrolyzation, carboxylation and etching with potassium hydroxide (KOH). For example, Ji et al. [98, 99] modified the SWNTs and MWNTs by etching using KOH and tested the etched CNTs for three pharmaceutical active compounds (i.e., sulfamethoxazole, tetracycline and tylosin). They found the adsorption performance of the KOH modified SWNTs (K-SWNTs) and KOH modified MWNTs (K-MWNTs) for sulfamethoxazole and tetracycline was enhanced by around 56% and 84% compared to the unetched SWNTs and MWNTs, respectively. This has been explained by increasing the surface area of the etched CNTs.

2.1.3 Graphene

Graphene is a two-dimensional carbonaceous nanomaterial formed from a layer of sp2 hybridized carbon atoms. The graphene nanomaterial has exceptional properties such as high specific surface area [98, 99], high electrocatalytic activity [100], great thermal conductivity [101], high stiffness and strength [102] and high speed electron mobility [103]. These unique physical properties attracted great interest of scientist and introduced it for different potential applications. Among these applications is the adsorptive removal of emerging pollutants such as pharmaceutical active compounds.

2.1.3.1 Types of graphene

The following are the common types of graphene:

- Single layer graphene (SLG): it is one thick hexagonally arranged sp2 hybridized bonded carbon atoms. The dimensions of SLG vary from nano- to microscale. It can be suspended in an aqueous solution or adhered on a substrate.

- Multilayer graphene (MLG): it consists of few flaks of single layer graphene and it is useful in the preparation of nanomaterial composites.

- Graphene oxide: it is a single layer or multilayer graphene that has high oxygenated surface and prepared by exfoliation and chemical oxidation of graphite.

- Reduced graphene oxide: it is that same as graphene oxide; however, the oxygenated functional groups are reduced chemically, thermally or biologically.

2.1.3.2 Graphene for pharmaceutical active compounds' removal

The graphene nanomaterials and their modified forms have extraordinary surface area and catalytic activity, and as a result, they can be used in several applications such as adsorptive removal of pharmaceutical active compounds [104–111]. Gao et al. [106] investigated the adsorption performance of graphene oxide for tetracycline antibiotic from aqueous solution. They found that the adsorption of tetracycline achieved mainly through a π-π and cation-π interactions with a maximum monolayer adsorption capacity is 313 mg/g and it decreased with an increase in the solution pH or the sodium ions concentration. In 2017, Danna et al. [107] modified a graphene oxide with decafluorobiphenyl and then investigated the prepared adsorbent for removal of six pharmaceutical active compounds from water namely, carbamazepine, sulfamethoxazole, sulfadiazine, ibuprofen, paracetamol and phenacetin. They found that the adsorption capacities for these compounds are 340.5 μmol/g, 428.3 μmol/g, 214.7 μmol/g, 224.3 μmol/g, 350.6 μmol/g and 316.1 μmol/g, respectively. A study in 2014 showed that the adsorptive removal of acetaminophen, aspirin and caffeine from aqueous solution using graphene nanoplates (GNPs) was thermodynamically spontaneous and exothermic with adsorption capacities of 18.07 mg/g, 12.98 mg/g and 19.72 mg/g for acetaminophen, aspirin and caffeine, respectively [105].

The surface area of graphene reduces significantly in solutions due to its aggregation, and as a result, the adsorption capacity of graphene is reduced, which is one of the main disadvantages associated with using graphene as adsorbents. Functionalization or modification of the graphene with certain functional group or metals can be the best solution to overcome that disadvantage as well as increase the adsorption capacity of graphene. Lin et al. [108] functionalized a graphene oxide with magnetic nanoparticles and then studied its adsorptive removal for four tetracycline (TC) pharmaceutical active compounds (i.e., tetracycline, oxytetracycline, chlortetracycline and doxycycline) from aqueous solution. They found that the solution pH and ionic strength had insignificant effect on the TC adsorption and the maximum adsorption capacity is 39.1 mg/g.

3. Conclusions

Pharmaceutical active compounds are continuously released into aquatic environment via different routes (i.e., human and animal excreta, medical industry's waste, wastewater effluent, sewage and landfill leaching). That release increases due to the increase of general use of pharmaceutical compounds in human and veterinary medicines. Therefore, these compounds should be removed from the contaminated water to prevent their accumulation, reduce the environmental pollution and provide an additional source of clean water. Removal of pharmaceutical active compounds from aquatic media can be achieved by either conventional or advanced methods. Among them, the adsorption technique has many advantages over the others. Several materials as adsorbents have been reported and discussed in the literature such as silica-based adsorbents, polymeric materials, clay, carbonaceous materials and other materials. Activated carbon, carbon nanotube and graphene oxide among carbonaceous materials show excellent performance and high adsorption capacity for pharmaceutical active compounds. As discussed in this chapter, the activated carbon can be activated using different methods (i.e., physical or chemical activation), while the carbon nanotube can be produced through using one of the following methods: (i) arc discharge, (ii) laser ablation and (iii) chemical vapor deposition. The physical (surface area and porosity) and chemical (functional groups) properties are significantly affected by the followed production method for these carbonaceous materials. Using freely available raw materials for the activated carbon and carbon nanotubes production and their modification with different nanoparticles and functional groups is the future prospect for the adsorptive removal of pharmaceutical active compounds from the aquatic environment.

Acknowledgements

The support of the Center for Environment and Water in the research institute of King Fahd University of Petroleum and Minerals is highly acknowledged.

Conflict of interest

The author declares that there are no conflicts of interest.

Author details

Mazen K. Nazal
Center for Environment and Water, Research Institute at king Fahd University for Petroleum and Minerals (KFUPM), Dhahran, Saudi Arabia

*Address all correspondence to: mazennazal@kfupm.edu.sa

References

[1] Ternes T. Occurrence of drugs in German sewage treatment plants and rivers. Water Research. 1998;**32**:3245-3260

[2] World Health Organization "Pharmaceuticals in Drinking-water" WHO/HSE/WSH/11.05. 2011

[3] Tabak HH, Bunch RL. Steroid hormones as water pollutants. I. Metabolism of natural and synthetic ovulation-inhibiting hormones by microorganisms of activated sludge and primary settled sewage. Developments in Industrial Microbiology. 1970;**11**:367-376

[4] Garrison AW, Pope JD, Allen FR. GC/MS analysis of organic compounds in domestic wastewaters. In: Keith LH, editor. Identification and Analysis of Organic Pollutants in Water. Ann Arbor, MI: Ann Arbor Science Publishers Inc.; 1976. pp. 517-556

[5] Hignite C, Azarnoff DL. Drugs and drug metabolites as environmental contaminants: Chlorophenoxyiso-butyrate and salicylic acid in sewage water effluent. Life Sciences. 1977;**20**(2):337-341

[6] Kolpin DW et al. Pharmaceuticals, hormones, and other organic wastewater contaminants in U.S. streams, 1999-2000: A national reconnaissance. Environmental Science & Technology. 2002;**36**:1202-1211

[7] Buser HR, Muller MD, Theobald N. Occurrence of the pharmaceutical drug clofibric acid and the herbicide mecoprop in various Swiss lakes and in the North Sea. Environmental Science & Technology. 1998;**32**(1):188-192

[8] Reddersen K, Heberer T, Dünnbier U. Identification and significance of phenazone drugs and their metabolites in ground- and drinking water. Chemosphere. 2002;**49**:539-544

[9] Bound A, Voulvoulis N. Household disposal of pharmaceuticals as a pathway for aquatic contamination in the United Kingdom. Environmental Health Perspectives. 2005;**113**:1705-1711

[10] DWI. "Desk based review of current knowledge on pharmaceuticals in drinking water and estimation of potential levels. Final Report Prepared by Watts and Crane Associates for Drinking Water Inspectorate, Department for Environment, Food and Rural Affairs (Defra Project Code: CSA 7184/WT02046/DWI70/2/213). 2007. Available from: http://dwi.defra.gov.uk/research/completed-research/reports/dwi70-2-213.pdf

[11] Ternes TA et al. Assessment of technologies for the removal of pharmaceuticals and personal care products in sewage and drinking water to improve the indirect potable water reuse. POSEIDON project detailed report (EU Contract No. EVK1-CT-2000-00047). 2005

[12] Bulloch DN et al. Occurrence of halogenated transformation products of selected pharmaceuticals and personal care products in secondary and tertiary treated wastewaters from Southern California. Environmental Science & Technology. 2015;**49**:2044-2051

[13] Bu Q, Wang B, Huang J, Deng S, Yu G. Pharmaceuticals and personal care products in the aquatic environment in China: A review. Journal of Hazardous Materials. 2013;**262**:189-211

[14] Jung C et al. Removal of endocrine disrupting compounds, pharmaceuticals, and personal care products in water using carbon nanotubes: A review. Journal of

Industrial and Engineering Chemistry. 2015;**27**:1-11

[15] Dong S et al. Recent developments in heterogeneous photocatalytic water treatment using visible light-responsive photocatalysts: A review. RSC Advances. 2015;**5**:14610-14630

[16] Richardson SD, Ternes TA. Water analysis: Emerging contaminants and current issues. Analytical Chemistry. 2014;**86**:2813-2848

[17] Gadipelly C, Pérez-González A, Yadav GD, Ortiz I, Ibáñez R, Rathod VK, et al. Pharmaceutical industry wastewater: Review of the technologies for water treatment and reuse. Industrial and Engineering Chemistry Research. 2014;**53**:11571-11592

[18] Wu X, Conkle JL, Ernst F, Gan J. Treated wastewater irrigation: Uptake of pharmaceutical and personal care products by common vegetables under field conditions. Environmental Science & Technology. 2014;**48**:11286-11293

[19] Miller EL, Nason SL, Karthikeyan KG, Pedersen JA. Root uptake of pharmaceutical and personal care product ingredients. Environmental Science & Technology. 2016;**50**:525-541

[20] Tanoue R et al. Uptake and tissue distribution of pharmaceuticals and personal care products in wild fish from treated-wastewater-impacted streams. Environmental Science & Technology. 2015;**49**:11649-11658

[21] Subedi B et al. Occurrence of pharmaceuticals and personal care products in German fish tissue: A national study. Environmental Science & Technology. 2012;**46**:9047-9054

[22] Rivera-Jiménez SM, Méndez-González S, Hernández-Maldonado A. Metal (M = Co^{2+}, Ni^{2+} and Cu^{2+}) grafted mesoporous SBA-15: Effect of transition metal incorporation and pH conditions on the adsorption of Naproxen from water. Microporous and Mesoporous Materials. 2010;**132**:470-479

[23] Rivera-Jiménez SM, Hernández-Maldonado AJ. Nickel (II) grafted MCM-41: A novel sorbent for the removal of naproxen from water. Microporous and Mesoporous Materials. 2008;**116**:246-252

[24] Esplugas S, Bila DM, Krause LGT, Dezotti M. Ozonation and advanced oxidation technologies to remove endocrine disrupting chemicals (EDCs) and pharmaceuticals and personal care products (PPCPs) in water effluents. Journal of Hazardous Materials. 2007;**149**:631-642

[25] Klavarioti M, Mantzavinos D, Kassinos D. Removal of residual pharmaceuticals from aqueous systems by advanced oxidation processes. Environment International. 2009;**35**:402-417

[26] Wang CJ, Li Z, Jiang WT, Jean JS, Liu CC. Cation exchange interaction between antibiotic ciprofloxacin and montmorillonite. Journal of Hazardous Materials. 2010;**183**:309-314

[27] Turku I, Sainio T, Paatero E. Thermodynamics of tetracycline adsorption on silica. Environmental Chemistry Letters. 2007;**5**:225-228

[28] Bao Khanh V, Shin EW, Snisarenko O, Jeong WS, Lee HS. Removal of the antibiotic tetracycline by Fe-impregnated SBA-15. Korean Journal of Chemical Engineer. 2010;**27**:116-120

[29] Punyapalakul P, Sitthisorn T. Removal of ciprofloxazin and carbamazepine by adsorption on functionalized mesoporous silicates. World Academy of Science, Engineering and Technology. 2010;**69**:546-550

[30] Lorphensri O, Intravijit J, Sabatini DA, Kibbey TC, Osathaphan K, Saiwan C. Sorption of acetaminophen, 17α-ethynyl estradiol, nalidixic acid, and norfloxacin to silica, alumina, and a hydrophobic medium. Water Research. 2006;**40**:1481-1491

[31] Robberson KA, Waghe AB, Sabatini DA, Butler EC. Adsorption of the quinolone antibiotic nalidixic acid onto anion-exchange and neutral polymers. Chemosphere. 2006;**63**:934-941

[32] María A, Otero RM, Rodrigues AE. Recovery of Vitamin B12 and cephalosporin-C from aqueous solutions by adsorption on non-ionic polymeric adsorbents. Separation and Purification Technology. 2004;**38**:85-98

[33] Scordino M, Di Mauro A, Passerini A, Maccarone E. Adsorption of flavonoids on resins: Hesperidin. Journal of Agricultural and Food Chemistry. 2003;**51**:6998-7004

[34] Chaubal MV, Payne GF, Reynolds CH, Albright RL. Equilibria for the adsorption of antibiotics onto neutral polymeric sorbents: Experimental and modeling studies. Biotechnology and Bioengineering. 1995;**47**:215-226

[35] Lee JW, Park HC, Moon H. Adsorption and desorption of cephalosporin c on nonionic polymeric sorbents. Separation and Purification Technology. 1997;**12**:1-11

[36] Dutta MK, Dutta NN, Bhattacharya KG. Aqueous phase adsorption of certain beta-lactam antibiotics onto polymeric resins and activated carbon. Separation and Purification Technology. 1999;**16**:213-224

[37] Ribeiro MH, Ribeiro IA. Modelling the adsorption kinetics of erythromycin onto neutral and anionic resins. Bioprocess and Biosystems Engineering. 2003;**26**:49-55

[38] KristiaPutra E, Pranowo R, Sunarso J, Indraswati N, Ismadji S. Performance of activated carbon and bentonite for adsorption of amoxicillin from wastewater: Mechanisms, isotherms and kinetics. Water Research. 2009;**43**:2419-2430

[39] Akçay G, Kılınç E, Akçay M. The equilibrium and kinetics studies of flurbiprofen adsorption onto tetrabutylammonium montmorillonite (TBAM). Colloids and Surfaces A. 2009;**335**:189-193

[40] Figueroa RA, Leonard A, MacKay AA. Modeling tetracycline antibiotic sorption to clays. Environmental Science & Technology. 2003;**38**:476-483

[41] Wang Y-J, Jia D-A, Sun R-J, Zhu H-W, Zhou D-M. Adsorption and cosorption of tetracycline and copper(II) on montmorillonite as affected by solution pH. Environmental Science & Technology. 2008;**42**:3254-3259

[42] Pils JRV, Laird DA. Sorption of tetracycline and chlortetracycline on K- and Ca-saturated soil clays, humic substances, and Clay–humic complexes. Environmental Science & Technology. 2007;**41**:1928-1933

[43] Bewick MWM. The adsorption and release of tylosin by clays and soils. Plant and Soil. 1979;**51**:363-372

[44] Zhang W, Ding Y, Boyd SA, Teppen BJ, Li H. Sorption and desorption of carbamazepine from water by smectite clays. Chemosphere. 2010;**81**:954-960

[45] Wu Q, Lia Z, Hong H, Yin K, Tie L. Adsorption and intercalation of ciprofloxacin on montmorillonite. Applied Clay Science. 2010;**50**:204-211

[46] Chang P-H, Jean J-S, Jiang W-T, Li Z. Mechanism of tetracycline sorption on rectorite. Colloids and Surfaces A. 2009;**339**:94-99

[47] Chang P-H, Li Z, Jiang W-T, Jean J-S. Adsorption and intercalation of tetracycline by swelling clay minerals. Applied Clay Science. 2009;**46**:27-36

[48] Martins AC, Pezoti O, Cazetta AL, Bedin KC, Yamazaki DAS, Bandoch GFG, et al. Removal of tetracycline by NaOH-activated carbon produced from macadamia nut shells: Kinetic and equilibrium studies. Chemical Engineering Journal. 2015;**260**:291-299

[49] Marzbali MH, Esmaieli M, Abolghasemi H, Marzbali MH. Tetracycline adsorption by H_3PO_4-activated carbon produced from apricot nut shells: A batch study. Process Safety and Environment Protection. 2016;**102**:700-709

[50] Acosta R, Fierro V, Martinez de Yuso A, Nabarlatz D, Celzard A. Tetracycline adsorption onto activated carbons produced by KOH activation of tyre pyrolysis char. Chemosphere. 2016;**149**:168-176

[51] Carl JR-U, Sánchez-PoloJesús VG-PM, Ocampo-Pérez JL-PR. Tetracycline removal from water by adsorption/bioadsorption on activated carbons and sludge-derived adsorbents. Journal of Environmental Management. 2013;**131**(15):16-24

[52] Rivera-Utrilla J, Prados-Joya G, Sánchez-Polo M, Ferro-García MA, Bautista-ToledoI.Removalofnitroimidazole antibiotics from aqueous solution by adsorption/bioadsorption on activated carbon. Journal of Hazardous Materials. 2009;**170**:298-305

[53] Çalışkan Salihi E, Göktürk S. Adsorption characteristics of sulfamethoxazole and metronidazole on activated carbon. Separation Science and Technology. 2010;**45**:244-255

[54] Oleszczuk P, Pan B, Xing B. Adsorption and desorption of oxytetracycline and carbamazepine by multiwalled carbon nanotubes. Environmental Science & Technology. 2009;**43**:9167-9173

[55] Mestre AS, Pires J, Nogueira JMF, Carvalho AP. Activated carbons for the adsorption of ibuprofen. Carbon. 2007;**45**:1979-1988

[56] Aksu Z, Tunç Ö. Application of biosorption for penicillin G removal: Comparison with activated carbon. Process Biochemistry. 2005;**40**:831-847

[57] Erdinç N, Göktürk S, Tunçay M. A study on the adsorption characteristics of an amphiphilic phenothiazine drug on activated charcoal in the presence of surfactants. Colloids and Surfaces B. 2010;**75**:194-203

[58] Ji L, Chen W, Zheng S, Xu Z, Zhu D. Adsorption of sulfonamide antibiotics to multiwalled carbon nanotubes. Langmuir. 2009;**25**:11608-11613

[59] Zimnitsky DS, Yurkshtovich TL, Bychkovsky PM. Adsorption of zwitterionic drugs on oxidized cellulose from aqueous and water/alcohol solutions. The Journal of Physical Chemistry. B. 2004;**108**:17812-17817

[60] Zhang H, Huang C-H. Adsorption and oxidation of fluoroquinolone antibacterial agents and structurally related amines with goethite. Chemosphere. 2007;**66**:1502-1512

[61] Wunder DB, Bosscher VA, Cok RC, Hozalski RM. Sorption of antibiotics to biofilm. Water Research. 2011;**45**:2270-2280

[62] Kikuta TUT. Separate estimation of adsorption and degradation

of pharmaceutical substances and estrogens in the activated sludge process. Water Research. 2005;**39**:1289-1300

[63] Chen W-R, Huang C-H. Adsorption and transformation of tetracycline antibiotics with aluminum oxide. Chemosphere. 2010;**79**:779-785

[64] Cheng G, Karthikeyan KG. Sorption of the antimicrobial ciprofloxacin to aluminum and iron hydrous oxides. Environmental Science & Technology. 2005;**39**:9166-9173

[65] Cheng G, Karthikeyan KG. Interaction of tetracycline with aluminum and iron hydrous oxides. Environmental Science & Technology. 2005;**39**:2660-2667

[66] Kahle M, Stamm C. Sorption of the veterinary antimicrobial sulfathiazole to organic materials of different origin. Environmental Science & Technology. 2006;**41**:132-138

[67] Feitosa-Felizzola J, Hanna K, Chiron S. Adsorption and transformation of selected human-used macrolide antibacterial agents with iron(III) and manganese (IV) oxides. Environmental Pollution. 2009;**157**:1317-1322

[68] Fang Z, Chen J, Qiu X, Qiu X, Cheng W, Zhu L. Effective removal of antibiotic metronidazole from water by nanoscale zero-valent iron particles. Desalination. 2011;**268**:60-67

[69] Caroni ALPF, de Lima CRM, Pereira MR, Fonseca JLC. The kinetics of adsorption of tetracycline on chitosan particles. Journal of Colloid and Interface Science. 2009;**340**:182-191

[70] Bui TX, Choi H. Adsorptive removal of selected pharmaceuticals by mesoporous silica SBA-15. Journal of Hazardous Materials. 2009;**168**:602-608

[71] Goyne KW, Chorover J, Kubicki JD, Zimmerman AR, Brantley SL. Sorption of the antibiotic ofloxacin to mesoporous and nonporous alumina and silica. Journal of Colloid and Interface Science. 2005;**283**:160-170

[72] Li Z, Ackley LSC, Fenske N. Adsorption of tetracycline on kaolinite with pH-dependent surface charges. Journal of Colloid and Interface Science. 2010;**351**:254-260

[73] Hamerlinck Y, Mertens DH. In: Vansant EF, editor. Activated Carbon Principles in Separation Technology. New York: Elsevier; 1994

[74] Mantell CL. Carbon and Graphite Handbook. New York: Interscience; 1968

[75] Mohan D, Singh KP. Granular activated carbon. In: Lehr J, Keeley J, Lehr J, editors. Water Encyclopedia: Domestic, Municipal, and Industrial Water Supply and Waste Disposal. New York: Wiley/Interscience; 2005

[76] Otowa T, Nojima Y, Miyazaki T. Development of KOH activated high surface area carbon and its application to drinking water purification. Carbon. 1997;**35**(9):1315-1339

[77] Mattson JS, Mark HB Jr. Activated Carbon. New York: Marcel Dekker; 1971

[78] Hower JC, Senior CL, Suuberg EM, Hurt RH, Wilcox JL, Olson ES. Mercury capture by native fly ash carbons in coal-fired power plants. Progress in Energy and Combustion Science. 2010;**36**:510-529

[79] Fletcher TH, Ma J, Rigby JR, Brown AL, Webb BW. Soot in coal combustion systems. Progress in Energy and Combustion Science. 1997;**23**(3):283-301

[80] Ahmed J. Adsorption of quinolone, tetracycline, and penicillin antibiotics

from aqueous solution using activated carbons: Review. Environmental Toxicology and Pharmacology. 2017;**50**:1-10

[81] Chen Y, Wang F, Duan L, Yang H, Gao J. Tetracycline adsorption onto rice husk ash, an agricultural waste: Its kinetic and thermodynamic studies. Journal of Molecular Liquids. 2016;**222**:487-494

[82] Craig MW. Coping with resistance to copper/silver disinfection. Water Engineering & Management. 2001;**148**(11):27

[83] Wang X, Li Q, Xie J, Jin Z, Wang J, Li Y, et al. Fabrication of ultralong and electrically uniform single-walled carbon nanotubes on clean substrates. Nano Letters. 2009;**9**(9):3137-3141. DOI: 10.1021/nl901260b

[84] Carbon Nanotube. 2020. Available from: http://en.wikipedia.org/wiki/Carbon_nanotube

[85] Kim, Li H. Fabrication and Applications of Carbon Nanotube-Based Hybrid Nanomaterials by Means of Non-Covalently Functionalized Carbon Nanotubes, Carbon Nanotubes—From Research to Applications. Stefano Bianco: IntechOpen; 2011

[86] Tasis D, Tagmatarchis N, Bianco A, Prato M. Chemistry of carbon nanotubes. Chemical Reviews. 2006;**106**:1105-1136

[87] Thoteson E, Ren Z, Chou T. Advances in the science and technology of carbon nanotubes and their composites: A review. Composites Science and Technology. 2001;**61**(13):1899-1912

[88] Zhidong H, Alberto F. Thermal conductivity of carbon nanotubes and their polymer nanocomposites: A review. Progress in Polymer Science. 2011;**36**(7):914-944

[89] Collins PG, Avouris P. Nanotubes for electronics. Scientific American. 2000;**283**:62-69

[90] Niyogi S, Hamon MA, Hu H, Zhao B, Bhowmik P, Sen R, et al. Chemistry of single-walled carbon nanotubes. Accounts of Chemical Research. 2002;**35**:1105-1113

[91] Hone J, Whitney M, Piskoti C, Zettl A. Thermal conductivity of single-walled carbon nanotubes. Physical Review B. 1999;**59**:R2514

[92] Ebbesen TW, Ajayan PM. Large-scale synthesis of carbon nanotubes. Nature. 1992;**358**:220-222

[93] The Smalley Group at Rice University. 2020. Available from: https://oit.rice.edu/

[94] Journet C, Bernier P. Production of carbon nanotubes. Applied Physics A: Materials Science & Processing. 1998;**67**(1):1-9

[95] Dai H. Carbon nanotubes: Opportunities and challenges. Surface Science. 2002;**500**(1-3):218-241

[96] Yu F, Sun S, Han S, Zheng J, Ma J. Adsorption removal of ciprofloxacin by multi-walled carbon nanotubes with different oxygen contents from aqueous solutions. Chemical Engineering Journal. 2016;**285**:588-595

[97] Yu F, Ma J, Han S. Adsorption of tetracycline from aqueous solutions onto multi-walled carbon nanotubes with different oxygen contents. Scientific Reports. 2014;**4**:5326

[98] Ji L, Shao Y, Xu Z, Zheng S, Zhu D. Adsorption of monoaromatic compounds and pharmaceutical antibiotics on carbon nanotubes activated by KOH etching. Environmental Science & Technology. 2010;**44**:6429-6436

[99] Stoller MD, Park S, Zhu Y, An J, Ruoff RS. Graphene-based ultracapacitors. Nano Letters. 2008;**8**:3498-3502

[100] He H, Gao C. Graphene nanosheets decorated with Pd, Pt, Au, and Ag nanoparticles: Synthesis, characterization, and catalysis applications. Science China Chemistry. 2011;**54**:397-404

[101] Balandin AA, Ghosh S, Bao W, Calizo I, Teweldebrhan D, Miao F, et al. Superior thermal conductivity of single-layer graphene. Nano Letters. 2008;**8**:902-907

[102] Lee C, Wei X, Kysar JW, Hone J. Measurement of the elastic properties and intrinsic strength of monolayer graphene. Science. 2008;**321**:385-388

[103] Novoselov KS, Geim AK, Morozov S, Jiang D, Zhang Y, Dubonos S, et al. Electric field effect in atomically thin carbon films. Science. 2004;**306**:666-669

[104] Sinha A, Jana NR. Graphene-based composite with γ-Fe_2O_3 nanoparticle for the high-performance removal of endocrine-disrupting compounds from water. Chemistry. 2012;**8**(4):786-791

[105] Al-Khateeb LA, Almotiry S, Salam MA. Adsorption of pharmaceutical pollutants onto graphene nanoplatelets. Chemical Engineering Journal. 2014;**248**:191-199

[106] Gao Y, Li Y, Zhang L, Huang H, Hu JJ, Shah SM, et al. Adsorption and removal of tetracycline antibiotics from aqueous solution by graphene oxide. Journal of Colloid and Interface Science. 2012;**368**(1):540-546

[107] Danna S, Shubo D, Jin L, Hubian W, Conghui H, Giovanni C, et al. Preparation of porous graphene oxide by chemically intercalating a rigid molecule for enhanced removal of typical pharmaceuticals. Carbon. 2017;**119**:101-109

[108] Lin Y, Xu S, Jia L. Fast and highly efficient tetracyclines removal from environmental waters by graphene oxide functionalized magnetic particles. Chemical Engineering Journal. 2013;**225**:679-685

[109] Kerkez-Kuyumcu O, Bayazit SS, Salam MA. Antibiotic amoxicillin removal from aqueous solution using magnetically modified graphene nanoplatelets. Journal of Industrial and Engineering Chemistry. 2016;**36**:198-205

[110] Kyzas GZ, Koltsakidou A, Nanaki SG, Bikiaris DN, Lambropoulou DA. Removal of beta-blockers from aqueous media by adsorption onto graphene oxide. Science of the Total Environment. 2015;**537**:411-420

[111] Kalantary RR, Azari A, Esrafili A, Yaghmaeian K, Moradi M, Sharafi K. The survey of Malathion removal using magnetic graphene oxide nanocomposite as a novel adsorbent: Thermodynamics, isotherms, and kinetic study. Desalination and Water Treatment. 2016;**57**(58):28460-28473

Chapter 6

Carbon-Based Materials (CBMs) for Determination and Remediation of Antimicrobials in Different Substrates: Wastewater and Infant Foods as Examples

Ahmed El-Gendy, Ahmed S. El-Shafie, Ahmed Issa, Saeed Al-Meer, Khalid Al-Saad and Marwa El-Azazy

Abstract

The widespread use of antimicrobials within either a therapeutic or a veterinary rehearsal has resulted in a crisis on the long run. New strains of antimicrobial-resistant microorganisms have appeared. Contamination of water with pharmaceutically active materials is becoming a fact! and efficacy of wastewater treatment plants is a question. Adsorption is a promising technique for wastewater treatment. Carbon-based materials are among the most commonly used adsorbents for remediation purposes. Food production and commercialization are posing rigorous regulations. In this concern, almost all authoritarian societies are setting up standards for the maximum residue levels permissible in raw and processed food. Among these products is infant foods. The current trend is to use carbon-based and recycled from agricultural wastes, which can selectively remove target antimicrobials. Nanoparticles are among the most commonly used materials. With the enormous amount of data generated from an analytical process, there is a need for a powerful data processing technique. Factorial designs play an important role in not only minimalizing the number of experimental runs, and hence saving chemicals, resources, and reducing waste but also, they serve to improve the sensitivity and selectivity, the most important analytical outcomes.

Keywords: pharmaceutically active materials (PhAMs), wastewater treatment, adsorption, carbon-based materials (CBMs), infant food, detection, factorial designs

1. Introduction

Drugs and pharmaceutically active materials (PhAMs) represent an enormous category of chemicals that include all materials with therapeutic effects (e.g., drugs that can be further classified according to their chemical structures, biological activities, and mechanism of action), cosmetics, supplementary and dietary products, personal care products (PCPs), X-rays contrast media, etc. Daily use of PhAMs is then becoming a fact. As per the Organization for Economic Co-operation and Development (OECD) report, in 2017, the expenditure on retail pharmaceuticals per capita was the highest in USA and averaged 564 $/person among the OECD countries [1].

It is noteworthy to mention that 75% of this amount was devoted to prescription drugs. With the increased awareness with health and health standards, the consumption of PhAMs is also escalating. As per FDA's (USA Food and Drug Administration) Center for Drug Evaluation and Research's (CDER) annual report, 59 novel drugs were approved in 2018, compared to 48 drugs in 2019. Of course, these approvals are associated with many new formulations being available for the consumer in the market [2, 3].

Representing a significant category of aquatic pollutants, PhAMs are usually released into the aquatic systems from different sources, including but not limited to: the effluents of the manufacturing sites and hospitals, illegal disposal, veterinary applications, and landfill leachate. The daily use by humans and the subsequent conversion of PhAMs into various metabolites with variable chemical structures is also a major source. The fate of these metabolites, and probably their parent drug compound, is usually the wastewater [4–8].

Antimicrobials (antibiotics, antifungals, antiseptics, antivirals, etc.) are also an enormous category of pharmaceuticals used mainly in the treatment and control of infectious diseases. Having unquestionable benefits for human and animal health, their use is becoming indispensable. However, *nothing is absolute!* The pervasive use of antimicrobials within either a therapeutic or a veterinary rehearsal (mainly antibiotics) especially in developing countries where medicine is usually dispensed as OTC (over the counter, non-prescription) drug has resulted in a crisis in the long run. Moreover, the excessive release of antibiotics into the surface and wastewater with ubiquitous concentrations reflects the magnitude of the problem [9]. New strains of antimicrobial-resistant microorganisms have appeared. These breeds are no longer responding to any medication, an issue that exaggerates the problem especially with the sluggish development of new drugs and therapeutics [10–13].

Figure 1 shows a schematic representation of the subcategories of antibiotics and their fate in the environment.

Antimicrobials can be further classified according to their chemical structures into subcategories; for example, antibiotics can be classified into subclasses such

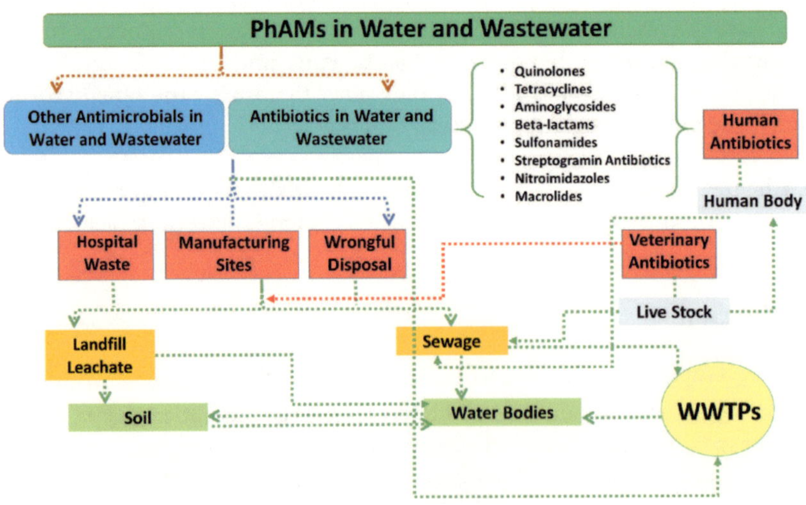

Figure 1.
Schematic representation of categories of antibiotics and their routes in the environment.

as quinolones and fluoroquinolones, tetracyclines, aminoglycosides, β-lactams, sulfonamides, streptogramin antibiotics, nitroimidazoles, and macrolides [14]. Due to this variable chemical structure, removal of this category of pollutants is an intricate task. Several approaches exist in literature for the remediation of wastewater samples from antimicrobials and their metabolites. Similarly, many techniques were described to detect anti-infective agents in foods and dietary supplements.

2. Chapter taxonomy

Throughout the following subsections, the discussion will be focused on applications of carbon-based materials (CBMs) for the detection of antimicrobials in food products, especially baby foods, as well as the remediation of wastewaters from antimicrobials. Though the aim is to survey the use of CBMs in the literature as a removal approach, a comparison between CBMs and other materials/techniques used will be useful in terms of evaluating their removal efficiency. Special attention will be paid to the methods conducted via the use of factorial designs and response surface methodological approaches.

3. Pharmaceutically active materials (PhAMs) in water and wastewater: frequency, fate, health risks, regulatory concerns, and removal

3.1 Occurrences and fate

As previously mentioned, PhAMs intimidate the water systems from different sources. Contaminated water bodies included wastewater as well as drinking water. Reported concentrations were in the range of parts-per-trillion (*ppt*) up to parts-per-billion (*ppb*), and included locations all over the world [15–21]. Higher concentrations, up to μg/L, were also reported for more than 160 drug species and PhAMs [9, 22]. In a comprehensive review prepared by Verlicchi et al. [16], the risk quotient (RQ) of 51 PhAMs was screened and assessed. Out of the investigated PhAMs, it was found that 14 compounds represent a high risk to the environment. The list included seven antibiotics from different subclasses (erythromycin, azithromycin, clarithromycin, tetracycline, ofloxacin, amoxicillin, and sulfamethoxazole), representing 50% of the high-risk PhAMs. The reminder of the list included antipsychotics, lipid-regulating, and anti-inflammatory drugs. Yet, 19 PhAMs were of a medium risk, and this time the list included eight antimicrobials (penicillin G, sulfadiazine, cefotaxime, enoxacin, trimethoprim, doxycycline, roxithromycin, and metronidazole), representing around 42% of the reported list. For antimicrobials, concentrations ranged between 0.001 and 32 μg/L, with the highest absolute and highest average concentrations being reported for ofloxacin and sulfadiazine/ofloxacin, respectively. Antifungals also had their share and clotrimazole was reported with a concentration of 0.029 μg/L [16].

The presence of these concentrations in wastewater as well as drinking water is raising several apprehensions about the competency of wastewater treatment plants (WWTPs) and the implemented remediation techniques. These concerns might be resolved when it is comprehended that residues of undegradable PhAMs reach the WWTPs through the urban wastewater compilation structures. The positive point is that some of these compounds might be completely degradable and can be effectually removed; however, others, possessing variable chemical structures with different physical, biological, and chemical properties, may not.

These compounds being widely variable in terms of hydrophobicity, polarity, solubility, volatility, absorbability, and binding abilities, as well as biodegradability can get through to the WWTPs even at very low concentrations [23–26]. The intricate task in removing PhAMs, therefore, stems from their properties. The majority of these materials possess acidic/basic functional groups, are of high polarity with high solubility, and cannot be easily degraded, or hydrolyzed. Yet, and in comparison to other contaminants (e.g., dyes/pigments, pesticides), PhAMs could be classified as *persistent* [27, 28].

Figure 1 shows a representation of the presence of antibiotics (human and veterinary) in the environment, and their routes until they reach the aquatic environments and WWTPs.

3.2 Health risks

The consequences of the existence of the PhAMs either in waste and drinking water or even in WWTPs is still indistinct. However, what is well understood is that the impact in the long run extends to human's and animal's health, the aquatic environment, and eventually the ecosystem. This effect is greatly dependent on the released dose of the PhAMs as well as their pharmacological effects. The issue becomes of concern when we know that the metabolites might be of a higher risk compared to the parent drug compound.

These effects and upon protracted exposure to PhAMs above the permissible limits include for example intoxication in human beings (e.g., the inundated enzyme-substrate relationship following long-term exposure to analgesics might result in elevated plasma concentrations and hence toxicity). Other adverse effects include somatic abnormalities, allergies, lung diseases, and hormonal disruption (e.g., in case of hormones, the adverse effects might instigate cancer following the destruction of DNA) [24, 29–31].

Moreover, the toxicity of the aquatic environment indirectly influences the human health. At the aquatic level, the metabolic mechanisms of aquatic microorganisms are affected. At the microbial level, microorganisms, upon prolonged exposure to anti-infectives, for example, become more tolerant and new strains, which cannot be cured using the conventional antimicrobials, are now in the scene. Influences include impaired reproduction, adversative effects on movement, and metabolism in mussels. Since PhAMs are not the only species released into the receiving aquatic bodies, adverse synergetic interactions between PhAMs and other contaminants should be expected [32–34].

3.3 Regulations and roles of regulatory bodies

The following few paragraphs will focus on the regulatory measures taken by the regulatory bodies and some countries to control the release of PhAMs, especially human drugs, in aquatic environments. Many countries have executed the environmental risk assessment (ERA) measures for the regulation of human PhAMs. Taking USA as an example, the National Environmental Policy Act of 1969 (NEPA) 21 C.F.R. 25.15, 40 C.F.R. 1508, necessitates all concerned organizations, for example, the U.S. EPA (Environmental Protection Agency) and the U.S. FDA, to implement the following measures:

- Assess the environmental influences of their actions,
- Perform an environmental assessment (EA),

- Create an environmental impact statement (EIS).

Consequently, any application to file a new drug to the US FDA requires the applicant to submit an EA or requires an entitlement for categorical exclusion. The latter would be approved in case the filed substance will not increase the concentration of an "active moiety"; or it will increase the concentration of this moiety, but the estimated concentration of this moiety as it reaches the aquatic body (entrance point) is below 1 ppb; or the filed material will neither change the distribution of the substance, nor its metabolites or degradation products [35]. In Canada, however, a priority substance list (PSL) has been founded and priority is given to the assessment of the toxicity of the PhAMs [36–38].

As per US FDA guidelines, and in order to estimate the expected introduction concentration (EIC) of a substance at the point of entry to the aquatic body, the following equation should be used:

$$\text{EIC–Aquatic (ppb)} = A * B * C * D \quad (1)$$

where,

A is the amount produced of the PhAM annually—as an active moiety—in kilograms and designated for direct use;

B is the L^{-1}/ day entering the publicly owned treatment works (POTWs);

C is the conversion factor (1 year/365 days); and

D is the conversion factor (10^9 μg/kg)

This equation was established assuming the following:

- No metabolism is taking place,
- All PhAMs manufactured annually are consumed and enter the POTWs,
- The use of the produced PhAMs happens thru the US and depends on the population and quantity of wastewater produced.

The European Union (EU), and on the other hand, has created two lists: one is a priority list, which includes 45 PhAMs where the environmental quality standards (EQS) have to be held in the highest regard for the disposal of these materials into the aquatic environments. The other list is a "watch-list" that compromises eight PhAMs, for which the risk to the aquatic environment needs to be investigated and verified [39]. The second list included antimicrobials, mainly antibiotics (azithromycin). Two other antibiotics, amoxicillin and ciprofloxacin, were also included in the same list but with different justifications. As per the commission, new "eco-toxicological" data were obtained for clarithromycin and azithromycin. **Table 1** shows the antimicrobials included in the watch-list, their analytical method of detection, and the maximum acceptable method detection limit (ng/L).

3.4 Water and wastewater treatment approaches

3.4.1 Classification of remediation approaches

Surveying the literature shows that different approaches have been reported for remediation of water and wastewater from PhAMs [24]. In general, wastewater treatment technologies can be categorized into *chemical* (e.g., ion exchange,

Name of the antimicrobial*	Indicative analytical method**,***	Maximum acceptable method detection limit (ng/L)
Macrolide antibiotics • *Erythromycin* • *Clarithromycin* • *Azithromycin*	SPE—LC-MS-MS	19
Metaflumizone	LLE—LC-MS-MS or SPE—LC-MS-MS	65
Amoxicillin	SPE—LC-MS-MS	78
Ciprofloxacin	SPE—LC-MS-MS	89

**CAS (Chemical Abstracts Service) and EU (European Union) numbers can be obtained from Ref. [30].*
***Extraction methods: LLE—liquid-liquid extraction; SPE—solid-phase extraction.*
****Analytical methods: GC–MS—Gas chromatography-mass spectrometry; LC-MS-MS—Liquid chromatography (tandem) triple quadrupole mass spectrometry.*

Table 1.
Some of the antimicrobials included in the watch-list of substances for Union-wide monitoring as set out in Article 8b of Directive 2008/105/EC.

chlorination, coagulation, ozonation, photo- and chemical oxidation); *physical* (e.g., membrane separation, and adsorption); and *biological* (e.g., biodegradation, membrane bioreactor, and enzyme bioreactor) approaches. Sometimes, combinational approaches are used as a series of treatment steps [24, 38–45]. Each of these approaches has its pros and cons. For example, chemical and physical approaches, unless being coupled to experimental design and response surface methodological approaches, would result in the generation of toxic byproducts, and the consumption of chemicals and solvents as well as time and resources. Overall, these techniques will not be green or ecofriendly. **Figure 2** shows a classification of the commonly used approaches in wastewater treatment.

3.4.2 Adsorption as a remediation approach

As previously mentioned, removal of PhAMs in WWTPs might be inadequate and occurs partially. Implementing the traditional remediation techniques, the removal of antibiotics was incomplete (e.g., removal of β-lactams was achieved with 17–43% efficiency, macrolides with 40–46% efficiency, sulfonamides with 66–90% efficiency, and tetracyclines with 66–90% efficiency) [46]. Adsorption as a physical/chemical remediation methodology is a versatile technique that has been extensively used in water and wastewater treatment from almost all types of contaminants (dyes, heavy metals, PhAMs, etc.) [6–8, 41, 43, 46–51].

As an approach, adsorption offers several advantages, mainly the availability of candidate adsorbents at almost no cost, easy application on the large scale with no toxic byproducts being generated, and most importantly, reasonable competency. Moreover, adsorption can be used as a removal approach following biological or chemical treatments in WWTPs. Several materials have been reported in literature as adsorbents. These adsorbents might be naturally occurring (e.g., carbon adsorbents recycled from agricultural waste products) [47–51], or synthetic (e.g., microporous and mesoporous carbons synthesized using Y zeolite and synthesized mesoporous silica as hard templates) [52]. Various adsorbents were reported in literature for the removal of antibiotics [53]; for example, carbonaceous materials [54], polymeric resins [55], chitosan [56], mesoporous materials [52], and molecularly imprinted polymers (MIPs) [57]. In the following subsections, the focus will

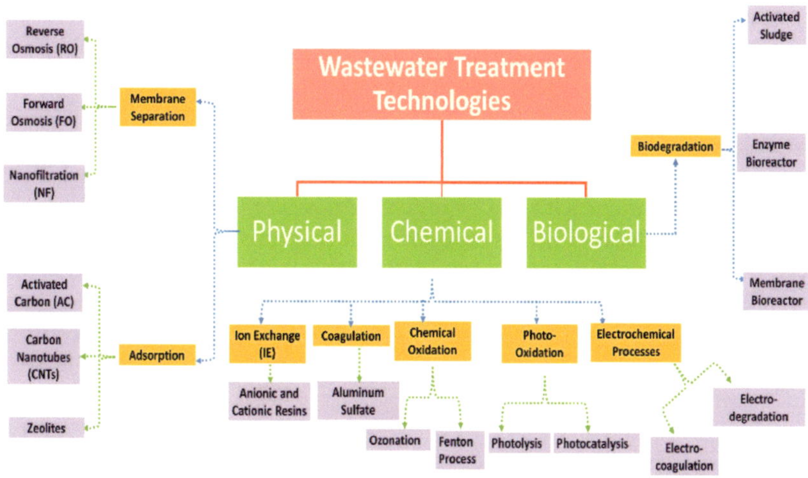

Figure 2.
Classification of wastewater treatment technologies.

be on carbon-based materials and their subcategories that are commonly used in the removal of antimicrobials.

3.4.3 Carbon-based materials (CBMs) for water and wastewater treatment

Possessing high surface area, pore size distribution, and pore volume, surface properties that enable strong interaction with the adsorbate, carbonaceous materials are among the most widely used and investigated adsorbents. In the following subsections, application of activated carbons (AC), graphene, carbon xerogels, and carbon nanotubes (CNTs), see **Figure 2**, in the treatment of water and wastewater from contaminants will be thoroughly discussed.

3.4.3.1 Activated carbons (ACs)

Activated carbon (AC) is generally a porous solid material with high surface area. With a high extent of microporosity, AC is a contemporaneous adsorbent that is widely used in water treatment on a large scale. According to the IUPAC classification, three types of materials can be recognized depending on the pore diameter: (1) microporous (pores < 2 nm diameter), (2) mesoporous (pores 2–50 nm diameter), and (3) macroporous (pores > 50 nm diameter) [58]. Activated carbon is an example of mesoporous materials, which is basically composed of a non-constant carbon framework. In reality, both micro- and mesopores exist in AC depending on the synthetic conditions.

As mentioned, AC can be obtained from agro-waste, carbonization at low temperature, and activation (physical "PA" or chemical "CA" depending on the temperature at which the activation/carbonization was conducted, and the experimental conditions where PA is usually conducted at anaerobic conditions compared to the use of inert gas in case of CA) [47–51, 53]. Adsorption capacity and efficiency are greatly dependent on the activation procedure followed and the nature of the raw material. Though CA takes place over a short period of time compared to PA, due to the usage of chemicals, CA might not be an ecofriendly approach [59, 60].

Treatment of AC using different approaches and with the purpose of increasing its adsorption efficiency for antimicrobials or even using it without activation has been the subject of lots of investigations. Treatment procedures included loading of magnetite nanoparticles, where for example, powdered date pits (DPs) were treated with a mixture of ferric and ferrous with a ratio of 2:1, followed by stirring at 60°C for 3 h, then neutralization using 4 M NaOH till pH 12. The mixture was washed with water several times, with methanol four times and then left to evaporate. **Figure 3** shows a picture for DPs before and after magnetization. Prepared AC was used for the removal of enrofloxacin and difloxacin from water samples. Other activation techniques included treatment with NaOH for the removal of tetracycline [61], H3PO4 and H4P2O7-AC for the removal of ciprofloxacin [62], composites for the removal of tetracycline [63], and sulfamethoxazole [64]. Untreated AC was also applied for the removal of antibiotics; for example, non-AC produced by pyrolysis of primary paper mill sludge was used for the removal of sulfamethoxazole [65] and tetracyclines [66].

3.4.3.2 Carbon nanotubes (CNTs)

First discovered in 1991, CNTs have become a target of hundreds of investigations. Representing an enormous transition in the field of nano-products, CNTs have seen an escalating interest in applications as well as investment. With extraordinary physicochemical properties, and feasibility of surface modification, CNTs are a *trap* for many environmental pollutants. Compared to the discrete structure of ACs, CNTs possess a more compact and well-defined structure. Three types of CNTs are now known: single-walled carbon nanotubes (SWCNTs), double-walled carbon nanotubes (DWCNTs), and multi-walled carbon nanotubes (MWCNTs). As their name implies, MWCNTs, unlike SWCNTs, are basically formed of cylinders arranged coaxially with the graphitic shells being alienated at a space of 0.34 nm, and a diameter of 1 nm. Therefore, the aspect ratio of the cylinder usually surpasses 10^5 and the CNTs are usually said to be highly directionally dependent or anisotropic. This feature, in addition to the noticeable chirality of the carbon atoms and their possible variable arrangements around the perimeter of the cylinder, the hollow structure with multiple adsorption sites, represents that of an *ideal* adsorbent [67–70]. **Figure 4** shows the possible adsorption sites in a CNT:

1. The external sites: located on the outside surface of each CNT;

2. The internal sites: the interior of individual CNT;

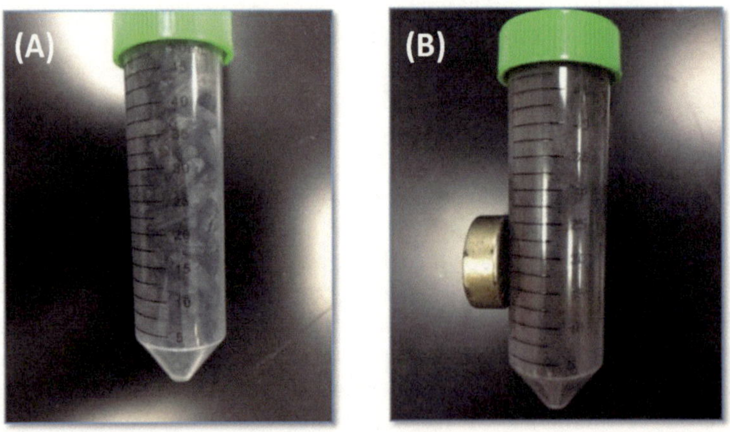

Figure 3.
Date pits (DPs) (A) before and (B) after loading of magnetite. Picture was taken in our laboratory at Qatar University.

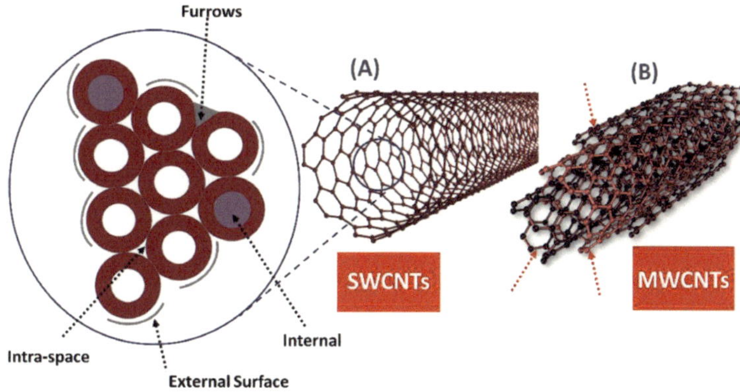

Figure 4.
(A) SWCNTs and (B) MWCNTs. The inset shows the possible adsorption sites.

3. The interstitial sites: located between the nanotubes;

4. The furrows: located across the peripheral intersection of two adjacent CNTs.

Applications of CNTs in the removal of antimicrobials has seen a major progress in the past few years. Untreated SWCNTs were used to remove tetracycline, sulfamethoxazole, and ciprofloxacin, oxytetracycline [71–74]. SWCNTs refluxed with 70% HNO_3 were used for the removal of triclosan [75]. Pristine MWCNTs were used for the removal of roxarsone [76]. Functionalization with carboxylic and hydroxyl moieties facilitated the removal of sulfapyridine [77].

3.4.3.3 Graphene-based materials

Graphene, graphene oxide (GO), and reduced graphene oxides (rGOs) are attracting a great deal of attention nowadays. Graphene is a 2D sheet in which the Sp^2-hybridized carbon atoms are arranged as a monolayer. With a large surface area, graphene can adsorb aquatic PhAMs via van der Waals or π-π electronic interactions [46, 78]. Since majority of antibiotics possess one or more cyclic component, they can be feasibly adsorbed by graphene via π-π interaction. Ciprofloxacin is among the antibiotics that have been removed using GS (graphene-soy gel) bio-composites [79]. GO was used to remove levofloxacin from aqueous solutions [80]. rGOs were similarly applied to remove sulfapyridine and sulfathiazole [81].

4. PhAMs in infant food: occurrence, health risks, regulations, and determination

4.1 Occurrence, health risks, and regulatory issues

Three categories of infant foods can be recognized: (1) infant formulas (0–6 months), (2) follow-on formulas (6–12 months), and (3) growth formulas (different ages after baby's first year and dependent on child's age). The composition of these formulas is variable and is dependent on the infant's age and nutritional needs of this age. Source of these formulas is therefore variable. Bovine milk is a major source for these formulas. Therefore, it is indispensable to ensure that these formulas are veterinary-drugs free!

The increasing understanding of the assembly of the food chain and the probability of infection of human with these resilient microorganisms either directly or via the food chain has explained largely the spread of these species. Therefore, the process of food production and commercialization is posing more rigorous regulations nowadays. In this regard, different societies, for example Food and Drug Administration (US FDA), European Union (EU), World Health Organization (WHO) in collaboration with Food and Agriculture Organization of the United Nations (FAO) creating the FAO/WHO Codex Alimentarius Commission (CAC), are setting up standards for the maximum residue levels (MRLs) permissible in raw and processed food products of animal or poultry origin. Yet, any food product that would conform to these criteria and the preceding risk assessments cannot be banned by countries of the World Trade Organization (WTO) [82–86].

Infant foods, in specific, should be monitored with a kind of scrupulousness either statutory or non-statutory. The main apprehension is that this food is to be offered to an age group that is the most susceptible for microbial infections and the chance of spread of resistant microorganisms becomes more likely. As per EU council regulation No. 2377/90, the MRL extends to include not only the intact drugs, but their degradation products as well as their metabolites. While the MRLs are well defined for a variety of baby foods, the situation is different for meat-, milk-, poultry-based infant foods, where the EU council is implementing the zero-tolerance policy; that is, the presence of such drugs in the said foods is totally banned. Existence of such a policy necessitates the presence of a sensitive analytical technique that can determine suspected drugs at even minute concentrations [86, 87].

4.2 Determination of antimicrobials in infant foods

Few methods exist in literature for the determination of antimicrobials in infant foods with major attention being directed to fluoroquinolones and tetracyclines, few antifungals, antiseptics, and antivirals. Techniques used ranged from pressurized liquid extraction followed by solid phase extraction (SPE) and LC-fluorescence detector analysis to ultra-high-performance liquid chromatography hyphenated to tandem mass spectrometry (UHPLC-MS/MS) and salting-out assisted liquid-liquid extraction (LLE) coupled to UHPLC-MS/MS [88–92].

With the progression in analytical method development, the current trend is to use miniaturized materials, which can selectively remove the target antimicrobial. Nanoparticles (NPs) either functionalized or non-functionalized, MWCNTs, molecularly imprinted polymers (MIPs), and graphene are among the most commonly used materials. Magnetic nanoparticles (MNPs) in particular and with their large surface area, and hence the swiftness of sorption, offer a great advantage in sample treatment [93, 94]. Surveying the literature shows that applications of CBMs in sample treatment are almost absent. In one of the investigations [95], the Zr-Fe-CMNPs composites were studied for sample pretreatment. It was shown that coating of the Zr-Fe MNPs with carbon increased elution efficiency of the studied fluoroquinolones, and therefore was used for the determination of fleroxacin, norfloxacin, and ofloxacin in meat-based baby food samples.

5. Multivariate analysis

Several parameters affect the adsorption of PhAMs either from wastewater or from foods. For example, parameters such as pH, contact time, adsorbent dose, initial adsorbate concentration, and ionic strength can affect adsorption efficiency of studied adsorbents [47–51].

Yet, very few of the techniques reported in literature implemented chemometrics or factorial designs as an approach and the regular one-variable-at-time (OVAT) approach is still used. In such an approach, only one variable is investigated at a time, with almost no idea about factorial interactions and no idea on how to deal with multiple-response variables concurrently. With the enormous amount of data generated from an analytical process, the need for a powerful data processing technique is needed. Chemometrics plays an important role not only in minimalizing the number of experimental runs, and hence saving chemicals, resources, and reducing waste but also in serving to improve the sensitivity and selectivity of the methods, the most important analytical outcomes [47–51, 96–99].

As a vision, Green Analytical Chemistry (GAC) adopts 12 principles that serve to compromise between the quality of an analytical process and the conservation of environment. Achieving such a settlement is an intricate task! Using chemometrics is one aspect of such an arrangement and ensuring a sustenance of the highest safety standards—both in water and wastewater remediation and in the production of infant foods—is another aspect.

Design of Experiments (DoE) as a multivariate approach is used to screen and then optimize the experimental conditions. The design usually entails two phases: screening (where all variables that might affect the process are investigated at wider levels), then optimization (where variables that were proved to be statistically significant from phase I are re-tested at narrower levels). Both phases are accompanied by statistical analysis using analysis of variance (ANOVA) [99].

In one of the investigations, a method based on pressurized liquid (PLE) and LC with fluorescence detection (LC-FLD) was used for the determination of residues of fluoroquinolones in baby foods. Factorial design was implemented in two phases. In the screening rehearsal, a fractional-factorial design was adopted to screen the impact of four parameters on the extraction process. Statistically significant variables as per ANOVA were further optimized using the face-centered central composite design [89, 100]. Applications of other designs were also reported [101].

6. Conclusions

The literature is rich with hundreds of articles that investigate the removal of antimicrobials from water and wastewater samples. Investigations that entail the usage of CBMs such as ACs, CNTs, and the graphene family, which possess unique physicochemical properties and most importantly a high surface area, are the most prevailing. Yet, and on the other hand, very few investigations on the determination of antimicrobials in baby foods, an important concern, are available in literature. Usage of CBMs in such a rehearsal is almost absent. All in all, removal of antimicrobials from wastewater and their determination in baby foods are usually affected by a number of variables. The common approach found in literature is the one based on the investigation of one-factor-at-a-time (OFAT). Application of chemometrics is still not as expected.

Acknowledgements

This work was made possible by Qatar University Internal Student Grant [QUCG-CAS-19/20-3] from Qatar University. The statements made herein are solely the responsibility of the authors.

Conflict of interest

The authors declare no conflict of interest.

Author details

Ahmed El-Gendy, Ahmed S. El-Shafie, Ahmed Issa, Saeed Al-Meer, Khalid Al-Saad and Marwa El-Azazy*
Department of Chemistry and Earth Sciences, College of Arts and Sciences, Qatar University, Doha, Qatar

*Address all correspondence to: marwasaid@qu.edu.qa

References

[1] OECD. Health at a Glance 2019: OECD Indicators, OECD Publishing, Paris [Internet]. 2019. Available from: https://doi.org/10.1787/4dd50c09-en

[2] Impact Innovation Predictability Access. FDA's Center for Drug Evaluation and Research [Internet]. 2020. Available from: https://www.fda.gov/drugs/new-drugs-fda-cders-new-molecular-entities-and-new-therapeutic-biological-products/new-drug-therapy-approvals-2019

[3] Rahmah AU, Harimurti S, Omar AA, Murugesan T. Optimization of oxytetracycline degradation inside UV/H2O2 reactor using box-Behnken experimental design. Journal of Applied Sciences. 2012;**12**:1154-1159. DOI: 10.3923/jas.2012.1154.1159

[4] Singh KP, Singh AK, Singh UV, Verma P. Optimizing removal of ibuprofen from water by magnetic nanocomposite using box-Behnken design. Environmental Science and Pollution Research. 2012;**19**:724-738. DOI: 10.1007/s11356-011-0611-4

[5] Aa N, Kommer GJ, Van Montfoort J, Versteegh J. Demographic projections of future pharmaceutical consumption in the Netherlands. Water Science and Technology. 2011;**63**:825-831. DOI: 10.2166/wst.2011.120

[6] Avcı A, Inci I, Baylan N. Adsorption of ciprofloxacin hydrochloride on multiwall carbon nanotube. Journal of Molecular Structure. 2020;**1206**:127711. DOI: 10.1016/j.molstruc.2020.127711

[7] Yu F, Sun S, Han S, Zheng J, Ma J. Adsorption removal of ciprofloxacin by multi-walled carbon nanotubes with different oxygen contents from aqueous solutions. Chemical Engineering Journal. 2016;**285**:588-595. DOI: 10.1016/j.cej.2015.10.39

[8] Carmosini N, Lee LS. Ciprofloxacin sorption by dissolved organic carbon from reference and bio-waste materials. Chemosphere. 2009;**77**:813-820. DOI: 10.1016/j.chemosphere.2009.08.003

[9] Larsson DGJ, de Pedro C, Paxeus N. Effluent from drug manufactures contains extremely high levels of pharmaceuticals. Journal of Hazardous Materials. 2007;**148**:751-755. DOI: 10.1016/j.jhazmat.2007.07.008

[10] Salmon GPC, Welch M. Antibiotic resistance: Adaptive evolution. The Lancet. 2008;**372**:S97-S103. DOI: 10.1016/S0140-6736(08)61888-7

[11] Kemper N. Veterinary antibiotics in the aquatic and terrestrial environment. Ecological Indicators. 2008;**8**:1-13. DOI: 10.1016/j.ecolind.2007.06.002

[12] WHO. Global strategy for containment of antimicrobial resistance, Executive summary 2001, World Health Organization (WHO), Geneva; 2001. Available from: https://www.who.int/drugresistance/WHO_Global_Strategy_English.pdf

[13] Meeting on Antimicrobial Resistance, Brussels. 2008. Available from: http://ec.europa.eu/food/resources/events en.htm

[14] Ullah H, Ali S. Classification of anti-bacterial agents and their functions. In: Kumavath RN, editor. Antibacterial Agents. IntechOpen; 2017. DOI: 10.5772/intechopen.68695. Available from: https://www.intechopen.com/books/antibacterial-agents/classification-of-anti-bacterial-agents-and-their-functions

[15] Gumbi BP, Moodley B, Birungi G, Ndungu PG. Detection and quantification of acidic drug residues in South African surface water using gas chromatography

mass spectrometry. Chemosphere. 2017;**168**:1042-1050. DOI: 10.1016/j.chemosphere.2016.10.105

[16] Verlicchi P, Al Aukidy M, Zambello E. Occurrence of pharmaceutical compounds in urban wastewater: Removal, mass load and environmental risk after a secondary treatment—A review. Science of the Total Environment. 2012;**429**:123-155. DOI: 10.1016/j.scitotenv.2012.04.028

[17] Sarmah AK, Meyer MT, Boxall ABA. A global perspective on the use, sales, exposure pathways, occurrence, fate and effects of veterinary antibiotics (VAs) in the environment. Chemosphere. 2006;**65**:725-759. DOI: 10.1016/j.chemosphere.2006.03.026

[18] Couto CF, Lange LC, Amaral MCS. Occurrence, fate and removal of pharmaceutically active compounds (PhACs) in water and wastewater treatment plants—A review. Journal of Water Process Engineering. 2019;**32**:100927. DOI: 10.1016/j.jwpe.2019.100927

[19] Valcarcel Y, Alonso SG, Rodríguez-Gil JL, Castano A, Montero JC, Criado-Alvarez JJ, et al. Seasonal variation of pharmaceutically active compounds in surface (Tagus River) and tap water (Central Spain). Environmental Science and Pollution Research. 2013;**20**:1396-1412. DOI: 10.1007/s11356-012-1099-2

[20] Watkinson AJ, Murby EJ, Kolpin DW, Costanzo SD. The occurrence of antibiotics in an urban watershed: From wastewater to drinking water. The Science of the Total Environment. 2009;**407**:2711-2723. DOI: 10.1016/j.scitotenv.2008.11.059

[21] Net S, Rabodonirin S, Sghaier RB, Dumoulin D, Chbi C, Tlili I, et al. Distribution of phthalates, pesticides and drug residues in the dissolved, particulate and sedimentary phases from transboundary rivers (France-Belgium). The Science of the Total Environment. 2015;**521**:152-159. DOI: 10.1016/j.scitotenv.2015.03.087

[22] Hernández F, Ibáñez M, Bade R, Bijlsma L, Sancho JV. Investigation of pharmaceuticals and illicit drugs in waters by liquid chromatography-high-resolution mass spectrometry. TrAC Trends in Analytical Chemistry. 2014;**63**:140-157. DOI: 10.1016/j.trac.2014.08.003

[23] Bell KY, Bandy J, Beck S, Keen O, Kolankowsky N, Parker AM, et al. Emerging pollutants – Part II: Treatment. Water Environment Research. 2012;**84**:1909-1940. DOI: 10.1002/wer.1233

[24] Taheran M, Brar SK, Verma M, Surampalli RY, Zhang TC, Valero JR. Membrane processes for removal of pharmaceutically active compounds (PhACs) from water and wastewaters. The Science of the Total Environment. 2016;**547**:60-77. DOI: 10.1016/j.scitotenv.2015.12.139

[25] Ziylan A, Ince NH. The occurrence and fate of anti-inflammatory and analgesic pharmaceuticals in sewage and fresh water: Treatability by conventional and non conventional processes. Journal of Hazardous Materials. 2011;**187**:24-37. DOI: 10.1016/j.jhazmat.2011.01.057

[26] Le Minh N, Khan SJ, Drewes JE, Stuetz RM. Fate of antibiotics during municipal water recycling treatment processes. Water Research. 2010;**44**:4295-4323. DOI: 10.1016/j.watres.2010.06.020

[27] Behera SK, Kim HW, Oh JE, Park HS. Occurrence and removal of antibiotics, hormones and several other pharmaceuticals in wastewater treatment plants of the largest industrial city of Korea. The Science of the Total Environment. 2011;**409**:4351-4360. DOI: 10.1016/j.scitotenv.2011.07.015

[28] Daneshkhah M, Hossaini H, Malakootian M. Removal of metoprolol from water by sepiolite-supported nanoscale zero-valent iron. Journal of Environmental Chemical Engineering. 2017;**5**:3490-3499. DOI: 10.1016/j.jece.2017.06.040

[29] Martin J, Camacho-Munoz D, Santos JL, Aparicio I, Alonso E. Occurrence of pharmaceutical compounds in wastewater and sludge from wastewater treatment plants: Removal and ecotoxicological impact of wastewater discharges and sludge disposal. Journal of Hazardous Materials. 2012;**239**:40-47. DOI: 10.1016/j.jhazmat.2012.04.068

[30] Khetan SK, Collins TJ. Human pharmaceuticals in the aquatic environment: A challenge to green chemistry. Chemical Reviews. 2007;**107**:2319-2364. DOI: 10.1021/cr020441w

[31] Yasuda MT, Sakakibara H, Shimoi K. Estrogen-and stress-induced DNA damage in breast cancer and chemoprevention with dietary flavonoid. Genes and Environment. 2017;**39**:1-9. DOI: 10.1186/s41021-016-0071-7

[32] Ebele AJ, Abdallah MA, Harrad S. Pharmaceuticals and personal care products (PPCPs) in the freshwater aquatic environment. Emerging Contaminant. 2017;**3**:1-16. DOI: 10.1016/j.emcon.2016.12.004

[33] Boyer EW. Management of opioid analgesic overdose. The New England Journal of Medicine. 2012;**367**:146-155. DOI: 10.1056/NEJMra1202561

[34] Carbajo JB, Petre AL, Rosal R, Herrera S, Leton P, Garcia-Calvo E, et al. Continuous ozonation treatment of ofloxacin: Transformation products, water matrix effect and aquatic toxicity. Journal of Hazardous Materials. 2015;**292**:34-43. DOI: 10.1016/j.jhazmat.2015.02.075

[35] U.S. FDA CDER/CBER. Guidance for Industry Environmental Assessment of Human Drug and Biologics Applications [Internet]. 1998. Available from: https://www.fda.gov/downloads/Drugs/Guidances/UCM444658.pdf [Accessed: 28 January 2020]

[36] Environment Canada. Overview of the Existing Substances Program [Internet]. 2007. Available from: https://www.ec.gc.ca/lcpe-cepa/715BA6D1-03F3-A724-B164-901EDE6723C3/OverviewOfESP_en.pdf [Accessed: 28 January 2020]

[37] Environment Canada. The Canadian Environmental Protection Act, 1999 and the Assessment of New Substances [Internet]. 2015b. Available from: https://www.canada.ca/en/environment-climatechange/services/canadian-environmental-protection-act-registry/general-information/fact-sheets/assessment-new-substances.html [Accessed: 28 January 2020]

[38] Geaniyu SO, Hullebusch EDV, Cretin M, Esposito G, Oturan MA. Coupling of membrane filtration and advanced oxidation processes for removal of pharmaceutical residues: A critical review. Separation and Purification Technology. 2015;**156** (Part 3):891-914. DOI: 10.1016/j.seppur.2015.09.059

[39] Commission Implementing Decision (EU), 2018/840 Of 5 June 2018 Establishing a Watch List of Substances for Union-wide Monitoring in the Field of Water Policy Pursuant to Directive 2008/105/EC of the European Parliament and of the Council and Repealing Commission Implementing Decision (EU) 2015/495 (notified Under Document C); 2018. p. 3362

[40] Madhavan J, Grieser F, Ashokkumar M. Combined advanced oxidation processes for the synergistic degradation of ibuprofen in aqueous environments. Journal of Hazardous

Materials. 2010;**178**:202-208. DOI: 10.1016/j.jhazmat.2010.01.064

[41] Pei M, Zhang B, He Y, Su J, Gin K, Lev O, et al. State of the art of tertiary treatment technologies for controlling antibiotic resistance in wastewater treatment plants. Environment International. 2019;**131**:105026. DOI: 10.1016/j.envint.2019.105026

[42] Bagheri H, Afkhami A, Noroozi A. Removal of pharmaceutical compounds from hospital wastewaters using nanomaterials: A review. Analytical and Bioanalytical Chemistry Research. 2016;**3**:1-18. DOI: 10.22036/abcr.2016.12655

[43] Carabineiro S, Thavorn-Amornsri T, Pereira M, Serp P, Figueiredo J. Comparison between activated carbon, carbon xerogel and carbon nanotubes for the adsorption of the antibiotic ciprofloxacin. Catalysis Today. 2012;**186**:29-34. DOI: 10.1016/j.cattod.2011.08.020

[44] Wang C-J, Li Z, Jiang W-T, Jean J-S, Liu C-C. Cation exchange interaction between antibiotic ciprofloxacin and montmorillonite. Journal of Hazardous Materials. 2010;**183**:309-314. DOI: 10.1016/j.jhazmat.2010.07.025

[45] Sun SP, Hatton TA, Chung T-S. Hyperbranched polyethyleneimine induced cross-linking of polyamide-imide nanofiltration hollow fiber membranes for effective removal of ciprofloxacin. Environmental Science & Technology. 2011;**45**:4003-4009. DOI: 10.1021/es200345q

[46] Peng B, Chen L, Que C, Yang K, Deng F, Deng X, et al. Adsorption of antibiotics on graphene and biochar in aqueous solutions induced by π-π interactions. Scientific Reports. 2016;**6**:31920. DOI: 10.1038/srep31920

[47] El-Azazy M, Dimassi S, El-Shafie AS, Issa AA. Bio-waste Aloe vera leaves as an efficient adsorbent for titan yellow from wastewater: Structuring of a novel adsorbent using Plackett-Burman factorial design. Applied Sciences. 2019;**9**:1-20. DOI: 10.3390/app9224856

[48] El-Azazy M, El-Shafie AS, Ashraf A, Issa AA. Eco-structured Biosorptive removal of basic Fuchsin using pistachio nutshells: A definitive screening design—Based approach. Applied Sciences. 2019;**9**:1-19. DOI: 10.3390/app9224855

[49] Al-Saad K, El-Azazy M, Issa AA, Al-Yafei A, El-Shafie AS, Al-Sulaiti M, et al. Recycling of date pits into a green adsorbent for removal of heavy metals: A fractional factorial design-based approach. Frontiers in Chemistry. 2019;**7-552**:1-16. DOI: 10.3389/fchem.2019.00552

[50] El-Azazy M, Kalla RN, Issa AA, Al-Sulaiti M, El-Shafie AS, Shomar B, et al. Pomegranate peels as versatile adsorbents for water purification: Application of box–Behnken design as a methodological optimization approach. Environmental Progress & Sustainable Energy. 2019;**38**:1-12. DOI: 10.1002/ep.13223

[51] El-Azazy M, El-Shafie AS, Issa AA, Al-Sulaiti M, Al-Yafie J, Shomar B, et al. Potato peels as an adsorbent for heavy metals from aqueous solutions: Eco-structuring of a green adsorbent operating Plackett-Burman design. Journal of Chemistry. 2019;**2019**:1-14. DOI: 10.1155/2019/4926240

[52] Ji L, Liu F, Xu Z, Zheng S, Zhu D. Adsorption of pharmaceutical antibiotics on template-synthesized ordered micro-and mesoporous carbons. Environmental Science & Technology. 2010;**44**:3116-3122. DOI: 10.1021/es903716s

[53] Yu F, Li Y, Han S, Ma J. Adsorptive removal of antibiotics from aqueous

solution using carbon materials. Chemosphere. 2016;**153**:365-385. DOI: 10.1016/j.chemosphere.2016.03.083

[54] Ahmed MJ, Theydan SK. Adsorption of cephalexin onto activated carbons from Albizia lebbeck seed pods by microwave-induced KOH and K2CO3 activations. Chemical Engineering Journal. 2012;**211**:200-207. DOI: 10.1016/j.cej.2012.09.089

[55] Dutta M, Dutta N, Bhattacharya K. Aqueous phase adsorption of certain β-lactam antibiotics onto polymeric resins and activated carbon. Separation and Purification Technology. 1999;**16**:213-224. DOI: 10.1016/S1383-5866(99)00011-8

[56] Adriano W, Veredas V, Santana C, Gonçalves L. Adsorption of amoxicillin on chitosan beads: Kinetics, equilibrium and validation of finite bath models. Biochemical Engineering Journal. 2005;**27**:132-137. DOI: 10.1016/j.bej.2005.08.010

[57] Xu L, Pan J, Dai J, Li X, Hang H, Cao Z, et al. Preparation of thermal-responsive magnetic molecularly imprinted polymers for selective removal of antibiotics from aqueous solution. Journal of Hazardous Materials. 2012;**233**:48-56. DOI: 10.1016/j.jhazmat.2012.06.056

[58] Rouquerol J, Avnir D, Fairbridge CW, Everett DH, Haynes JM, Pernicone N, et al. Recommendations for the characterization of porous solids (technical report). Pure and Applied Chemistry. 1994;**66**:1739-1758

[59] Wigmans T. Industrial aspects of production and use of activated carbons. Carbon. 1989;**27**:13-22. DOI: 10.1016/0008-6223(89)90152-8

[60] Holden M. Manufacture and uses of activated carbon. Effluent & Water Treatment Journal. 1982;**22**:27-46

[61] Martins AC, Pezoti O, Cazetta AL, Bedin KC, Yamazaki DA, Bandoch GF, et al. Removal of tetracycline by NaOH activated carbon produced from macadamia nut shells: Kinetic and equilibrium studies. Chemical Engineering Journal. 2015;**260**:291-299. DOI: 10.1016/j.cej.2014.09.017

[62] Sun Y, Yue Q, Gao B, Huang L, Xu X, Li Q. Comparative study on characterization and adsorption properties of activated carbons with H3PO4 and H4P2O7 activation employing Cyperus alternifolius as precursor. Chemical Engineering Journal. 2012a;**181**:790-797. DOI: 10.1016/j.cej.2011.11.098

[63] Shao L, Ren Z, Zhang G, Chen L. Facile synthesis, characterization of a MnFe2O4/activated carbon magnetic composite and its effectiveness in tetracycline removal. Materials Chemistry and Physics. 2012;**135**:16-24. DOI: 10.1016/j.matchemphys.2012.03.035

[64] Akhtar J, Amin NS, Aris A. Combined adsorption and catalytic ozonation for removal of sulfamethoxazole using Fe2O3/CeO2 loaded activated carbon. Chemical Engineering Journal. 2011;**170**:136-144. DOI: 10.1016/j.cej.2011.03.043

[65] Calisto V, Ferreira CI, Oliveira JA, Otero M, Esteves VI. Adsorptive removal of pharmaceuticals from water by commercial and waste-based carbons. Journal of Environmental Management. 2015;**152**:83-90. DOI: 10.1016/j.jenvman.2015.01.019

[66] Choi KJ, Kim SG, Kim SH. Removal of tetracycline and sulfonamide classes of antibiotic compound by powdered activated carbon. Environmental Technology. 2008;**29**:333-342. DOI: 10.1080/09593330802102223

[67] Iijima S. Helical microtubules of graphitic carbon. Nature. 1991;**354**:56-58. DOI: 10.1038/354056a0

[68] Pan B, Xing BS. Adsorption mechanisms of organic chemicals on carbon nanotubes. Environmental Science & Technology. 2008;**42**:9005-9013. DOI: 10.1021/es801777n

[69] Iijima S, Ichihashi T. Single-shell carbon nanotubes of 1-nm diameter. Nature. 1993;**363**:603-605. DOI: 10.1038/363603a0

[70] Rena X, Chena C, Nagatsu M, Wang X. Carbon nanotubes as adsorbents in environmental pollution management: A review. Chemical Engineering Journal. 2011;**170**:395-410. DOI: 10.1016/j.cej.2010.08.045

[71] Ji L, Chen W, Duan L, Zhu D. Mechanisms for strong adsorption of tetracycline to carbon nanotubes: A comparative study using activated carbon and graphite as adsorbents. Environmental Science & Technology. 2009;**43**:2322-2327. DOI: 10.1021/es803268b

[72] Kim H, Hwang YS, Sharma VK. Adsorption of antibiotics and iopromide onto single-walled and multi-walled carbon nanotubes. Chemical Engineering Journal. 2014;**255**:23-27. DOI: 10.1016/j.cej.2014.06.035

[73] Peng H, Pa B, Wu M, Liu Y, Zhang D, Xing B. Adsorption of ofloxacin and norfloxacin on carbon nanotubes: Hydrophobicity-and structure-controlled process. Journal of Hazardous Materials. 2012;**233**:89-96. DOI: 10.1016/j.jhazmat.2012.06.058

[74] Ncibi MC, Sillanpää M. Optimized removal of antibiotic drugs from aqueous solutions using single, double and multi-walled carbon nanotubes. Journal of Hazardous Materials. 2015;**298**:102-110. DOI: 10.1016/j.jhazmat.2015.05.025

[75] Cho H-H, Huang H, Schwab K. Effects of solution chemistry on the adsorption of ibuprofen and triclosan onto carbon nanotubes. Langmuir. 2011;**27**:12960-12967. DOI: 10.1021/la202459g

[76] Hu J, Tong Z, Hu Z, Chen G, Chen T. Adsorption of roxarsone from aqueous solution by multi-walled carbon nanotubes. Journal of Colloid and Interface Science. 2012;**377**:355-361. DOI: 10.1016/j.jcis.2012.03.064

[77] Tian Y, Gao B, Morales VL, Chen H, Wang Y, Li H. Removal of sulfamethoxazole and sulfapyridine by carbon nanotubes in fixed-bed columns. Chemosphere. 2013;**90**:2597-2605. DOI: 10.1016/j.chemosphere.2012.11.010

[78] Zhuang Y, Yu F, Ma J, Chen J. Graphene as a template and structural scaffold for the synthesis of a 3D porous bio-adsorbent to remove antibiotics from water. RSC Advances. 2015;**5**:27964-27969. DOI: 10.1039/c4ra12413h

[79] Zhuang Y, Yu F, Ma J, Chen J. Adsorption of ciprofloxacin onto graphene-soy protein biocomposites. New Journal of Chemistry. 2015;**39**:3333-3336. DOI: 10.1039/C5NJ00019J

[80] Dong S, Sun Y, Wu J, Wu B, Creamer AE, Gao B. Graphene oxide as filter media to remove levofloxacin and lead from aqueous solution. Chemosphere. 2016;**150**:759-764. DOI: 10.1016/j.chemosphere.2015.11.075

[81] Liu FF, Zhao J, Wang S, Xing B. Adsorption of sulfonamides on reduced graphene oxides as affected by pH and dissolved organic matter. Environmental Pollution. 2016;**210**:85-93. DOI: 10.1016/j.envpol.2015.11.053

[82] WHO. Worldwide strategy for the containment of antimicrobial resistance, WHO 2001. Available from: http://whqlibdoc.who.int/hq/2001/WHO_CDS_CSR_DRS_2001.2a.pdf

[83] USDA. Foreign Agricultural Service. Veterinary drug MRL

database. Available from: http://www.mrldatabase.com/default.cfm?selectvetdrug=1

[84] FAO/WHO. Available from: http://www.codexalimentarius.net/mrls/vetdrugs/jsp/vetd_q-s.jsp

[85] Motarjemi Y, Lelieveld H. Fundamentals in management of food safety in the industrial setting: Challenges and outlook of the 21st century. In: Motarjemi Y, Lelieveld H, editors. Food Safety Management: A Practical Guide for the Food Industry. Academic Press; 2014. DOI: 10.1016/B978-0-12-381504-0.00001-9. Available from: https://vdocuments.site/food-safety-management-a-practical-guide-for-the-food-industry-gnv64.html

[86] Commission Regulation 1353/2007/EC of 20th November, amending Council Regulation 2377/90/EC laying down a Community procedure for the establishment of maximum residue limits of veterinary medicinal products in foodstuffs of animal origin. Official Journal, L 303

[87] Hereber T, Lahrssen-Wiederholt M, Schafft H, Abraham K, Pzyrembel H, Henning KJ, et al. Zero tolerances in food and animal feed- are there any scientific alternatives? A European point of view of an international controversy. Toxicology Letters. 2007;**175**:118-135. DOI: 10.1016/j.toxlet.2007.10.002

[88] Díaz-Álvarez M, Turiel E, Martín-Esteban A. Selective simple preparation for the analysis of (fluoro) quinolones in baby food: Molecularly imprinted polymers versus anion exchange resins. Analytical and Bioanalytical Chemistry. 2009;**393**:899-905. DOI: 10.1007/s00216-008-2300-9

[89] Rodriguez E, Navarro Villoslada F, Moreno-Bondi MC, Marazuela MD. Optimization of a pressurized liquid extraction method by experimental design methodologies for the determination of fluoroquinolone residues in infant foods by liquid chromatography. Journal of Chromatography. A. 2010;**1217**:605-613. DOI: 10.1016/j.chroma.2009.11.089

[90] Gentili A, Perret D, Marchese S, Sergi M, Olmi C, Curini R. Accelerated solvent extraction and confirmatory analysis of sulphonamide residues in raw meat and infant foods by liquid chromatography electrospray tandem mass spectrometry. Journal of Agricultural and Food Chemistry. 2004;**52**:4614-4624. DOI: 10.1021/jf0495690

[91] Aguilera-Luiz MM, Martínez Vidal JL, Romero-González R, Garrido Frenich A. Multiclass method for fast determination of veterinary drug residues in baby food by ultra-high-performance liquid chromatography–tandem mass spectrometry. Food Chemistry. 2012;**132**:2171-2180. DOI: 10.1016/j.foodchem.2011.12.042

[92] Moreno-González D, García-Campaña AM. Salting-out assisted liquid–liquid extraction coupled to ultra-high-performance liquid chromatography–tandem mass spectrometry for the determination of tetracycline residues in infant foods. Food Chemistry. 2017;**221**:1763-1769. DOI: 10.1016/j.foodchem.2016.10.107

[93] He X, Wang GN, Yang K, Liu HZ, Wu XJ, Wang JP. Magnetic graphene dispersive solid phase extraction combining high performance liquid chromatography for determination of fluoroquinolones in foods. Food Chemistry. 2017;**221**:1226-1231. DOI: 10.1016/j.foodchem.2016.11.035

[94] Xu J-J, An M, Yang R, Tan Z, Hao J, Cao J, et al. Determination of tetracycline antibiotic residues in honey and milk by miniaturized solid phase extraction using chitosan-modified graphitized multi-walled carbon nanotubes. Journal of Agricultural and

Food Chemistry. 2016;**64**:2647-2654. DOI: 10.1021/acs.jafc.6b00748

[95] Vakh C, Alaboud M, Lebedinets S, Korolev D, Postnov V, Moskvin L, et al. An automated magnetic dispersive micro-solid phase extraction in a fluidized reactor for the determination of fluoroquinolones in baby food samples. Analytica Chimica Acta. 2018;**1001**:59-69. DOI: 10.1016/j.aca.2017.11.065

[96] Elazazy MS, Issa AA, Al-Mashreky M, Al-Sulaiti M, Al-Saad K. Application of fractional factorial design for green synthesis of cyano-modified silica nanoparticles: Chemometrics and multifarious response optimization. Advanced Powder Technology. 2018;**29**:1204-1215. DOI: 10.1016/j.apt.2018.02.012

[97] Elazazy MS, Ganesh K, Sivakumar V, Huessein YHA. Interaction of p-synephrine with p-chloranil: Experimental design and multiple response optimization. RSC Advances. 2016;**6**:64967-64976. DOI: 10.1039/C6RA10533E

[98] Elazazy MS. Determination of midodrine hydrochloride via Hantzsch condensation reaction: A factorial design based spectrophotometric approach. RSC Advances. 2015;**5**:48474-48483

[99] Elazazy MS. Factorial design and machine learning strategies: Impacts on pharmaceutical analysis. In: Zafar F, editor. Spectroscopic Analyses. IntechOpen; 2017. DOI: 10.5772/intechopen.69891. Available from: https://www.intechopen.com/books/spectroscopic-analyses-developments-and-applications/factorial-design-and-machine-learning-strategies-impacts-on-pharmaceutical analysis

[100] Jia W, Chu X, Ling Y, Huang J, Chang J. High-throughput screening of pesticide and veterinary drug residues in baby food by liquid chromatography coupled to quadrupole Orbitrap mass spectrometry. Journal of Chromatography. A. 2014;**1347**:122-128. DOI: 10.1016/j.chroma.2014.04.081

[101] Zhang B, Han X, Gu P, Fang S, Bai J. Response surface methodology approach for optimization of ciprofloxacin adsorption using activated carbon derived from the residue of desilicated rice husk. Journal of Molecular Liquids. 2017;**238**:316-325. DOI: 10.1016/j.molliq.2017.04.022

Section 2

Eco-Friendly Approach to Carbon and Carbon Materials

Chapter 7

Nanoporous Carbon Composites for Water Remediation

Benoît Cagnon, Marius Sebastian Secula and Şahika Sena Bayazit

Abstract

Metal–organic frameworks (MOFs) are known for their superior surface properties such as surface area and porosity. Thermal decomposition of MOFs may lead to nanoporous carbon composites. These composites can be further used in various application areas. Environmental remediation is one of the most popular areas for using these composites. Nowadays, nanoporous carbon composites are used generally in supercapacitors, lithium-ion batteries, and sensors. Besides the aforementioned application areas, these materials can be used as adsorbents, photocatalysts, and nanomotors. In this review, the preparation methods of nanoporous carbon materials will be explained and their use in environmental remediation will be summarized. The future perspectives of nanoporous carbon composites will be also discussed.

Keywords: metal-organic frameworks, nanoporous carbon composites, water remediation, organic pollutants

1. Introduction

Carbon finds its place in every new developing technological field. Carbon materials are used as drug delivery agent in biotechnological and pharmaceutical researches. They are used as electrode materials for developing new battery systems. Also, different kinds of carbon materials are used in sensor applications. Environmental engineering is another field that makes use of carbon materials as adsorbents or catalysts.

Nanoporous carbon materials are critically important for environmental remediation processes. The physicochemical properties of nanoporous carbon materials are especially suitable for water treatment operations. The specific surfaces area and porosity of carbon materials are very high [1]. Furthermore, different functional groups can be grafted to the surface of carbonaceous materials. Carbon materials are of a wide variety. Activated carbons, carbon nanotubes, graphene, and graphite can be mentioned as some important examples. Particularly, activated carbons derived from organic materials have been mostly used in water purification processes [2, 3].

In the last decade, metal-organic frameworks (MOFs) and their carbon derivatives have triggered strong interest from researchers. MOFs are formed by the combination of metal ions or clusters and organic linkers. The ranges of porosity and surface area of MOFs are quite wide compared to those of the other materials used for environmental remediation. The crystal motifs of MOFs are coordinated as periodically organometallic 3D structures [4]. The carbon structures derived from MOFs are prepared by pyrolysis method. Before the pyrolysis step, physical or chemical treatments are applied to the organic precursors [5]. The diversity of MOFs provides the potential to achieve numerous different kinds of nanoporous carbon materials. These materials can be used for different application areas.

The purpose of this chapter is to introduce the reader in the field of MOF-derived nanoporous carbon materials and their applications to environmental engineering. Firstly, MOF structures, preparation methods, and properties will be defined, and then nanoporous carbon composites (NCCs) will be presented. Additionally, the application areas of NCCs will be reviewed.

2. Metal-organic frameworks (MOFs)

Metal-organic frameworks are built by the combination of organic and inorganic sites. Several examples of MOFs are given in **Figure 1** (the structures were drawn using Mercury 4.2.0 software [6, 7]).

As shown in **Figure 1**, metal ions are connected by organic linkers. Different kinds of organic linkers can be used for preparation of MOFs. Terephthalic acid, trimesic acid, fumaric acid, and oxalic acid are only a few examples of organic linkers. A wide range of MOFs can be produced with a large number of different metal-organic sequences. MOFs have some advanced properties compared to traditional porous materials. Flexibility, regular structure, and ability of design are some of the most important properties. The crystalline structure of MOFs facilitates the characterization steps. X-ray diffraction analysis proves the structure of MOFs [11]. MOFs can be modified during the synthesis step and/or after synthesis. Synthesis media of MOFs can be easily controlled, so that some modifications can be operated by simply changing the media. For example, acid modulation is a preferred method for defecting MOFs and obtaining higher porosity. Also, organic linkers can be used to functionalize MOFs [11]. Nitroterephthalic acid and 2-aminoterephtalic acid are used for preparation of modified MOFs.

MOFs have a wide range of application areas. Gas storage and gas adsorption are very well-known application areas of MOFs. Also, sensors, catalysts, drug delivery agents, and biotechnological applications in general can be listed as applications of MOFs [12].

MOFs and MOF-based technologies have been started to be used as commercial products. The number of patents granted by different countries has been increasing in the last decade [13]. Different application types can be generated based on MOFs.

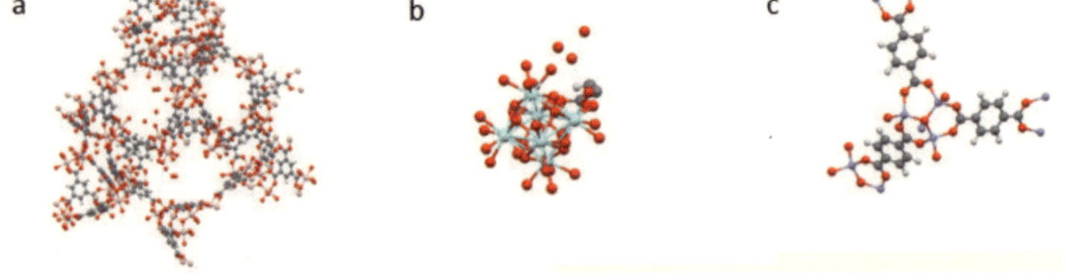

Figure 1.
(a) MIL-100 (Al) [8], (b) UiO-66 (Zr) [9], and (c) MOF-5 (Zn) [10].

MOFs can be transformed further to nanoporous carbon structures. Pyrolyzed MOFs have gained special attention in the last decades. This chapter focuses on NCCs. Preparation methods and application areas of NCCs are summarized.

3. Nanoporous carbon composites (NCC) derived from MOFs

Porous nanomaterials represent one of the most important classes of solid materials. Especially, nanoporous carbons have advanced chemical and physical features. Thermal and chemical stabilities of nanoporous carbons are remarkable so that these materials are very important for a wide range of technological areas. Many researchers reported that nanoporous carbon materials are very important for solving of the global environmental issues [1]. Nanoporous carbons are generally prepared by pyrolysis of an organic material [5].

Nanoporous carbons are very light, and they have high surface area and porosity. These properties represent important advantages [1]. Different organic materials can be used as precursor of nanoporous carbons. Generally, some templates are used in pyrolysis for obtaining ordered porous carbons [14]. Hard template and soft template methods have been used to prepare ordered nanoporous carbons. However, the stability of porous carbons prepared by hard template method is not appropriate. Also, this method is not economic and the process is very long. Soft template method is often not preferred by researchers due to it is difficult to find a suitable template for pyrolysis [14]. MOFs are newly used precursors for this purpose. The nanoporous carbons derived from MOFs provide different functionalities to the final product. Magnetic properties, oxygen and nitrogen doping, and metal or metal oxides on nanoporous carbons are some examples of functionalities [15]. Examples of application areas for nanoporous carbon composites are as follows: electrode materials, catalysts, and adsorbents [15].

The research studies on the topic of NCCs have been rapidly increasing in the last decade. As shown in **Figure 2**, the research activities have been increasing exponentially in the last 5 years.

According to the pyrolysis method, three different types of MOF structures can be obtained. **Figure 3** describes the three structures. Carbons and different types of metal compounds can be prepared by pyrolysis of MOFs under different conditions.

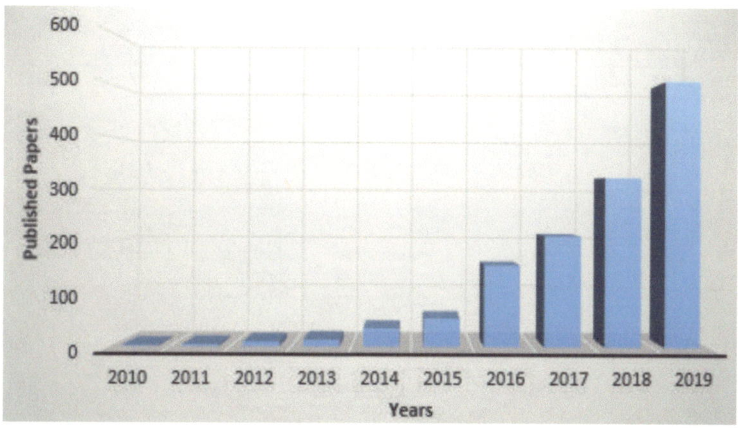

Figure 2.
The number of published papers about MOF-derived carbon composites during 2010–2019 (The data of graph was obtained from Scopus using MOF-derived carbon as keyword in January 5, 2020).

Metal carbides, metal oxides, and metal sulfides are some of the metal structures. These materials are generally used as heterogeneous catalysts [16].

The pyrolysis temperature and post- or pretreatments on MOFs can affect the obtained products. Some examples are given below.

As listed in **Table 1**, reaction conditions are very important for the final product. Furfuryl alcohol treated MOF-5 has different values of surface area in dependence of temperature. For IRMOF-1, IRMOF-3, and IRMOF-8, the total pore volume of MOFs increases. The total pore volume of IRMOF-1 increases from 1.45 to 4.06 cm^3/g; similarly, the total pore volume of IRMOF-3 increases from 0.90 to 2.01 cm^3/g [1]. Different carbon sources and different templates can be added to MOFs so that a wide variety of materials can be produced. Furfuryl alcohol is an example for this purpose. Zhang et al. [17] used glucose+ZIF-7 for preparing carbon structures. Glucose is a green carbon source and also ensures the removal of metals from the derived carbon structure. The additives for MOFs and derived products from these types of structures can be classified as follows (**Figure 4**).

MOFs have metal ions in their crystal structures. Direct pyrolysis under N_2 or Ar atmosphere cannot remove all metal compounds. According to the required structure,

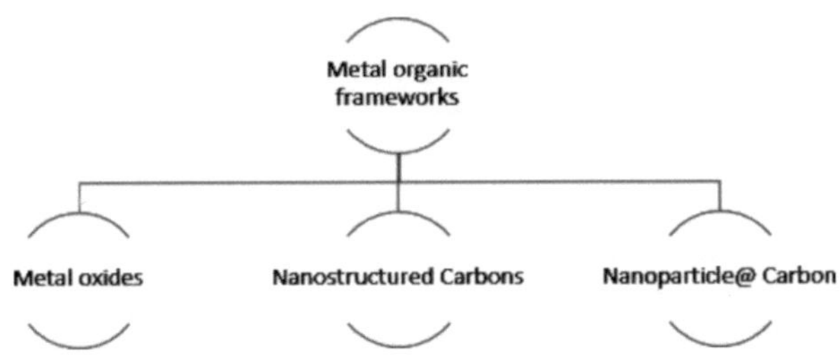

Figure 3.
The obtained product type after MOF pyrolysis [17].

MOFs	BET surface area (m^2/g)	Pyrolysis conditions	Derived structure	BET surface area (m^2/g)	Reference
Furfuryl alcohol treated MOF-5	-	Carbonization at 1000°C in Ar	Nanoporous carbon	2872	[5]
Furfuryl alcohol treated MOF-5	-	Carbonization at 800°C in Ar	Nanoporous carbon	417	[5]
Furfuryl alcohol treated MOF-5	-	Carbonization at 530°C in Ar	Nanoporous carbon	217	[5]
IRMOF-1	3447	Carbonization at 900°C in Ar	Nanoporous carbon	3174	[1]
IRMOF-3	2351	Carbonization at 900°C in Ar	Nanoporous carbon	1678	[1]
IRMOF-8	1735	Carbonization at 900°C in Ar	Nanoporous carbon	1978	[1]
MAF-4	968	Carbonization at 800°C in N_2	Nanoporous carbon	1176	[15]
MAF-5	454	Carbonization at 800°C in N_2	Nanoporous carbon	938	[15]
MAF-6	1339	Carbonization at 800°C in N_2	Nanoporous carbon	1642	[15]
$CO_3(NDC)_3(DMF)_4$	-	Calcination at 600°C in air	Co_3O_4	5.3	[18]

Table 1.
Examples of MOF-derived carbon composites.

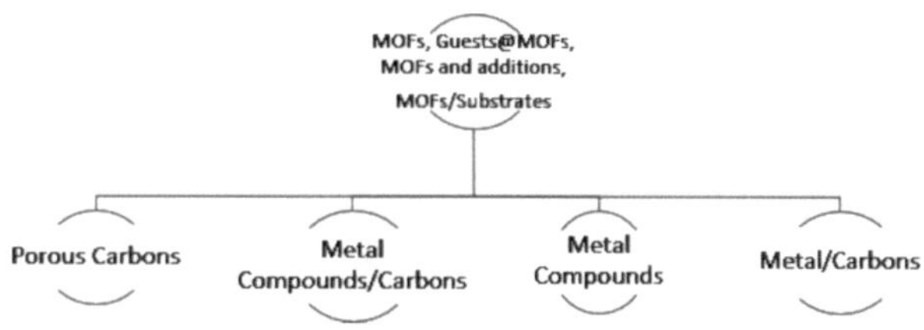

Figure 4.
The classification of products derived from different types of MOF composites [16].

MOFs are modified before pyrolysis. For this purpose, different templates can be used. Xia et al. [18] prepared a catalyst for oxygen reduction reaction. They used a Co-based MOF for this purpose to prepare Co@Co_3O_4@C core@bishell nanoparticles. In the mentioned study, the authors used MOF as a metal source, obtaining highly porous carbon structure. Also, they prepared Co@Co_3O_4 by controlled oxidation.

The ordered metal-free nanoporous carbon structures can be prepared by acid leaching [19] or by adding different materials such as glucose, etc. Graphene, carbon nanotubes, and other organic templates are used as extra carbon sources. Carbon sources provide larger surface areas for the required product. The surface area of furfuryl alcohol+MOF-5 can be given as an example.

The facile production of MOFs and MOF composites and also of nanoporous carbon composites derived from MOFs is very important for researchers. This is the easiest way for producing new catalysts and adsorbents. In the last 5 years, the researchers have prepared new derived materials and generally used them as catalysts in the energy field. In the present review, the uses of these derived materials in the area of water remediation are presented.

4. Application areas of nanoporous carbon composites

Electrochemical and energy areas are generally the most preferred applications for nanoporous carbon composites. The research group of Prof. Yusuke Yamauchi is one of the most active groups on this subject. For instance, they synthesized Zn^{2+}/Co^{2+} hybrid MOFs at room temperature, using deionized water as solvent [20]. After the pyrolysis of the hybrid material and activation with KOH (for increasing the surface area), they used the derived material for electrical double-layer capacitors and pseudocapacitors [20]. The nanoporous carbon composites are also used in oxygen reduction reaction [21], selective fluorescence sensor [22], and Li-ion battery applications [23]. Yamauchi et al. [24] prepared ZIF-8 nanoparticles and transformed it to nanoporous carbon to use as intracellular drug (cisplatin) delivery carriers.

Water remediation studies of MOF-derived nanoporous carbon composites are presented herein further.

Water is the most essential source of life for all living things. The greater part of water is salty, so a low percent of water is actually reachable and useful for livings. Even less, about 0.3% of water in the world can be used for human being [25]. Reaching clean water and cleaning the polluted water represent the biggest problem of the entire world. Carbon materials have been used for remediation of polluted water for a long time. New engineered materials are very promising. MOF-derived nanoporous carbon composites are one of these engineered materials. Some of these materials have started to be used in the last decade especially in the energy field. However, they have very huge potential for water remediation applications. The researchers show more and more interest to this purpose [5].

Different kinds of MOF-derived nanoporous carbon composites have been reported in water remediation studies. Some of these studies are introduced and listed below.

Li et al. [26] prepared nanoporous carbon from the derivation of MOF-5. The surface area value of the nanoporous carbon is 1731 m^2/g. They used the obtained material as an adsorbent for the removal of **sulfamethoxazole**, **bisphenol A**, and **methyl orange** from water. The maximum adsorption capacities (q_m) of these organic pollutants are 625 mg/g, 757 mg/g, and 872 mg/g, respectively. These values show that the nanoporous carbon composites are really effective for such organic pollutants. As the authors stated "these q_m values are 1.0–3.2 times higher than single-walled carbon nanotubes (SWCNTs) and powder active carbon (PAC) under the same conditions"; the reported results are truly remarkable.

Bhadra et al. [27] used metal azolate frameworks for the preparation of nanoporous carbons. After 24 h of thermal treatment, the surface area of nanoporous carbon reached 1906 m^2/g. The adsorption studies were carried out to remove aromatic hydrocarbons such as **naphthalene**, **anthracene**, **pyrene**, and **benzene** by means of this carbon. The q_m values are 150 mg/g for benzene, 240 mg/g for naphthalene, 280 mg/g for anthracene, and 310 mg/g for pyrene.

Zhang et al. [28] chose a bimetallic ZIF for the preparation of nanoporous carbon composites. ZIF (Zn/Co) was used as the precursor. The thermal conditions were 900°C, the atmosphere was Argon, and the reaction duration was of 5 h. The obtained surface area is of 398 m^2/g. This material was used for **Rhodamine B** adsorption. The maximum adsorption capacity was found as 116.2 mg/g. This value is competitive compared to the results in literature.

Xu et al. [29] used ZIF-8 and a carbon sources to prepare nanoporous carbon. The carbon sources were sucrose and dicyanamide. Carbonization conditions were 950°C under Ar atmosphere, and the obtained product was nitrogen-doped nanoporous carbon composites. The chosen adsorbate for this carbon was **methylene blue**. The highest value for the surface area is of 1796 m^2/g with dicyanamide treated ZIF-8. The q_m value is 1160.5 mg/g. According to literature, this value is outstanding.

Jin et al. [30] synthesized (Ni^{2+}/Zn^{2+}) + H3BTC MOFs to prepare further nanoporous carbon composites. Ni/Zn MOF was calcinated at 910°C under nitrogen flow. The surface area was measured as 999 m^2/g. The adsorption of **malachite green**, **Congo red**, **rhodamine B**, **methylene blue**, and **methyl orange** was carried out. Maximum adsorption capacities of organic dyes were found as 898, 818, 395, 312, and 271 mg/g, respectively.

Bhadra and Jhung [15] prepared MAF-4-, MAF-5-, and MAF-6-derived nanoporous carbon composites for the removal of emerging contaminants from water. The carbonization conditions were 800°C, under nitrogen flow for a 6-h period. The surface areas of carbonized products were found as 1176, 938, and 1642 m^2/g, respectively. The chosen contaminants are **salicylic acid**, **clofibric acid**, **diclofenac sodium**, **bisphenol A**, and **oxybenzone**. In the case of oxybenzone, the adsorption capacity of carbonized MAF-4 was 260 mg/g, 240 mg/g for carbonized MAF-5, and 440 mg/g for carbonized MAF-6.

Liu et al. [31] chose ZIF-8 (Zn) for carbonization. They added another carbon source to their MOFs. Furfuryl alcohol is generally used for carbon sources in such studies. The heat treatment temperature was 900°C under N_2 flow. They used the obtained adsorbent for dispersive solid phase extraction of **benzoylurea insecticides (diflubenzuron, triflumuron, hexaflumuron and teflubenzuron)**. The importance of this study resided in the very low pollution concentration range, 0.10–0.23 ng/L, the reported carbon getting very successful adsorption values.

Mg-MOF-74 is another template for using carbonization studies. Lv et al. [32] carried out the calcination at 800°C for 1 hour, under inert conditions. The product was

MgO/C composite. The surface area of Mg-MOF-74 was found as 18 m^2/g, yet the carbonized product has a significantly higher surface area, 296 m^2/g. The aim of producing this composite consisted in **uranium (VI)** adsorption. The adsorption capacity of the carbonized product was found as 777 mg/g, whereas the adsorption capacity of raw Mg-MOF-74 was of only 110 mg/g. In the literature, U (VI) adsorption capacity of activated carbon was given as 28 mg/g. This result outlines the huge potential of engineered nanoporous carbon composites for water remediation studies.

The carbonization process provides additional physical properties to carbon composites. Magnetization is one of them. Ahsan et al. [33] carbonized Fe-benzenedicarboxylate MOFs at 800°C under Ar for 1 h. The obtained product was magnetic C/Fe composite. They used this composite for **4-nitrophenol** and **methyl orange** degradation. Reaction yields reached 99.59% for 4-nitrophenol and 98% for methyl orange. The composites can be regenerated and used at least four times for this purpose. Lin et al. [34] prepared magnetic composite by carbonization of ZIF-67 and used this composite for **caffeine** reduction.

Bhadra et al. [35] carbonized and annealed Bio-MOF-1 (adenine, biphenyl-4,4′-dicarboxylic acid, and Zn^{2+}) at 1000°C (N_2 atmosphere) for different values of reaction time (6, 12, and 24 h). After 12 h, the maximum surface area was of 1449 m^2/g. **Bisphenol A** adsorption was carried out with this composite. The maximum adsorption capacity was found as 710 mg/g using 12-h annealed composite.

Torad et al. [36] prepared magnetic cobalt nanoparticles (Co/nanoporous carbon composites) by the carbonization of ZIF-67 crystals. The obtained Co nanoparticles were used for **methylene blue** adsorption. Two different values of temperature were chosen for carbonization, 600°C and 800°C, under N_2 flow. The authors also used HCl solutions for controlling the magnetization. The surface area of the MOF carbonized at 800°C was measured at 345 m^2/g. The maximum adsorption capacity is of 502.5 mg/g.

Fe(III)-modified MOF-5 was prepared by Chen et al. [37] The carbonization temperature was 500°C in the reported study. The product was used for the adsorption of **atrazine**, **carbamazepine**, **bisphenol A**, **norfloxacin**, and **4-nitrophenol**.

Sarker et al. [38] designed different MOFs for carbonization. They prepared ZIF-8 + ionic liquid composites. Then, the composite was carbonized at 1000°C, under nitrogen flow for 10 h. The surface area of this pyrolysis product is of 1468 m^2/g. The chosen adsorbents were two toxic herbicides **diuron** and **2,4-D.** The q_m values were found as 284 mg/g for diuron and 448 mg/g for 2,4-D. Bhadra et al. [39] adsorbed **ibuprofen** and **diclofenac** by ZIF-8-derived nanoporous carbon. Li et al. [40] also chose ZIF-8 for carbonization and adsorbed **ciprofloxacin** antibiotic from water. They showed a very good adsorption capacity of 416.7 mg/g.

In this section, some applications of MOF-derived nanoporous carbons in the field of water remediation were described. Generally, organic pollutants were chosen by researchers for adsorption studies. Only a few heavy metal adsorption studies were reported in literature. Organic dyes, personal care products, some drugs, and antibiotics were studied with only a few different MOFs. The investigations show that these materials have huge research potential for water remediation studies.

5. Conclusion

The properties of metal-organic frameworks and MOF-derived nanoporous carbon composites were discussed in this mini-review chapter. The structures of MOFs were explained. The preparation methods and the kinds of nanoporous carbons were mentioned.

The carbonization of MOFs represents the main method for preparing nanoporous carbon composites. The chosen temperature, the reaction media, and the reaction time are very important for the obtained product. MOFs can be calcinated under air atmosphere. Metal and metal oxides are the products. If inert atmosphere (N_2, Ar) is chosen, then carbon, carbon/metal, and carbon/metal oxides nanoparticles can be obtained. Also, some carbon sources can be used to provide higher surface areas, such as furfuryl alcohol, glucose, and sucrose. Some other templates can be used such as carbon nanotube, graphene, and other graphene structures. Different shapes and sizes of products can be synthesized. According to the chosen metal ions, the obtained nanoporous carbon composites can have magnetic properties. This property is very useful for adsorption processes.

The nanoporous carbons and carbon composites have been used for water remediation studies especially in the last 5 years. Generally, organic pollutants have been studied until now. Some organic dyes, personal care products, some drugs, herbicides, and antibiotics have been adsorbed from aqueous solutions using nanoporous carbon composites. According to the reported results, the adsorption efficiency of nanoporous carbon composites is very promising.

Having excellent physical and chemical properties, MOF-derived nanoporous carbon composites may lead to the development of countless different products. All these new products can be functionalized by various chemicals according to requirements. This new family of nanoporous materials provides large perspectives especially in the field of wastewater treatment. The summary presented herein above shows that highly featured MOF-derived composites have not been adequately used for water treatment so far. In the context of to the nowadays global problem of clean water scarcity, the future of MOF-derived nanoporous carbon composites emerges bright.

6. Future perspectives

The nanoporous carbon composites derived from MOFs are brand new materials. The surface area and porosities are very high. Such properties favor these carbon materials for adsorption and catalysis processes. Generally, energy applications and sensors have been studied with these carbon materials. Water remediation approaches based on nanoporous carbon composites are very promising.

Preparation methods of nanoporous carbons, diversifying the MOF structures before carbonization, enrich the research options. Further, these options open the perspectives for new products and new application areas. Water treatment activities will be positively influenced by this rich range of products. Engineered nanoporous carbon composites may be a significant milestone in solving the global environmental problems.

Conflict of interest

The authors declare no conflict of interest.

Author details

Benoît Cagnon[1], Marius Sebastian Secula[2] and Şahika Sena Bayazit[3]*

1 ICMN – Interfaces Confinement Matériaux Nanostructures, UMR 7374-CNRS, Université d'Orléans, Orléans Cedex 2, France

2 TUIaşi – Gheorghe Asachi Technical University of Iasi, Iasi, Romania

3 Mechanical Engineering Department, Beykent University, Sarıyer, Istanbul, Turkey

*Address all correspondence to: sahikasena@gmail.com

References

[1] Yang SJ, Kim T, Im JH, et al. MOF-derived hierarchically porous carbon with exceptional porosity and hydrogen storage capacity. Chemistry of Materials. 2012;**24**:464-470

[2] Wong S, Ngadi N, Inuwa IM, et al. Recent advances in applications of activated carbon from biowaste for wastewater treatment: A short review. Journal of Cleaner Production. 2018;**175**:361-375

[3] Hameed BH, Salman JM, Ahmad AL. Adsorption isotherm and kinetic modeling of 2,4-D pesticide on activated carbon derived from date stones. Journal of Hazardous Materials. 2009;**163**:121-126

[4] Reinoso DM, Diaz U, Frechero MA. Structural study of functional hierarchical porous carbon synthesized from metal-organic framework template. Materials Today Chemistry. 2019;**14**:1-10. DOI: 10.1016/j.mtchem.2019.08.007

[5] Chaikittisilp W, Ariga K, Yamauchi Y. A new family of carbon materials: Synthesis of MOF-derived nanoporous carbons and their promising applications. Journal of Materials Chemistry A. 2013;**1**:14-19

[6] Macrae CF, Bruno IJ, Chisholm JA, et al. Mercury CSD 2.0—New features for the visualization and investigation of crystal structures. Journal of Applied Crystallography. 2008;**41**:466-470

[7] Macrae CF, Edgington PR, McCabe P, et al. Mercury: Visualization and analysis of crystal structures. Journal of Applied Crystallography. 2006;**39**:453-457

[8] Volkringer C, Popov D, Loiseau T, et al. Synthesis, single-crystal X-ray microdiffraction, and NMR characterizations of the giant pore metal-organic framework aluminum trimesate MIL-100. Chemistry of Materials. 2009;**21**:5695-5697

[9] Øien S, Wragg D, Reinsch H, et al. Detailed structure analysis of atomic positions and defects in zirconium metal–organic frameworks. Crystal Growth & Design. 2014;**14**:5370-5372

[10] Trousselet F, Archereau A, Boutin A, et al. Heterometallic metal–organic frameworks of MOF-5 and UiO-66 families: Insight from computational chemistry. Journal of Physical Chemistry C. 2016;**120**:24885-24894

[11] Li J-R, Sculley J, Zhou H-C. Metal–organic frameworks for separations. Chemical Reviews. 2012;**112**:869-932

[12] Efome JE, Rana D, Matsuura T, et al. Insight studies on metal-organic framework nanofibrous membrane adsorption and activation for heavy metal ions removal from aqueous solution. ACS Applied Materials & Interfaces. 2018;**10**:18619-18629

[13] Yap MH, Fow KL, Chen GZ. Synthesis and applications of MOF-derived porous nanostructures. Green Energy & Environment. 2017;**2**:218-245

[14] Shen K, Chen X, Chen J, et al. Development of MOF-derived carbon-based nanomaterials for efficient catalysis. ACS Catalysis. 2016;**6**:5887-5903

[15] Bhadra BN, Jhung SH. A remarkable adsorbent for removal of contaminants of emerging concern from water: Porous carbon derived from metal azolate framework-6. Journal of Hazardous Materials. 2017;**340**:179-188

[16] Chen YZ, Zhang R, Jiao L, et al. Metal–organic framework-derived porous materials for catalysis.

Coordination Chemistry Reviews. 2018;**362**:1-23

[17] Oar-Arteta L, Wezendonk T, Sun X, et al. Metal organic frameworks as precursors for the manufacture of advanced catalytic materials. Materials Chemistry Frontiers. 2017;**1**:1709-1745

[18] Liu B, Zhang X, Shioyama H, et al. Converting cobalt oxide subunits in cobalt metal-organic framework into agglomerated Co_3O_4 nanoparticles as an electrode material for lithium ion battery. Journal of Power Sources. 2010;**195**:857-861. DOI: 10.1016/j.jpowsour.2009.08.058

[19] Marpaung F, Kim M, Khan JH, et al. Metal–organic framework (MOF)-derived nanoporous carbon materials. Chemistry – An Asian Journal. 2019;**14**:1331-1343

[20] Marpaung F, Park T, Kim M, et al. Gram-scale synthesis of bimetallic ZIFs and their thermal conversion to nanoporous carbon materials. Nanomaterials. 2019;**9**:1796

[21] Pandiaraj S, Aiyappa HB, Banerjee R, et al. Post modification of MOF derived carbon via g-C_3N_4 entrapment for an efficient metal-free oxygen reduction reaction. Chemical Communications. 2014;**50**:3363-3366

[22] Xu H, Zhou S, Xiao L, et al. Fabrication of a nitrogen-doped graphene quantum dot from MOF-derived porous carbon and its application for highly selective fluorescence detection of Fe^{3+}. Journal of Materials Chemistry C. 2015;**3**:291-297

[23] Banerjee A, Upadhyay KK, Puthusseri D, et al. MOF-derived crumpled-sheet-assembled perforated carbon cuboids as highly effective cathode active materials for ultra-high energy density Li-ion hybrid electrochemical capacitors (Li-HECs). Nanoscale. 2014;**6**:4387-4394

[24] Torad NL, Li Y, Ishihara S, et al. MOF-derived NANOPOROUS carbon as intracellular drug delivery carriers. Chemistry Letters. 2014;**43**:717-719

[25] Bhattacharya S, Gupta AB, Gupta A, et al. Introduction to Water Remediation: Importance and Methods. Singapore: Springer. 2018. pp. 3-8

[26] Li X, Yuan H, Quan X, et al. Effective adsorption of sulfamethoxazole, bisphenol A and methyl orange on nanoporous carbon derived from metal-organic frameworks. Journal of Environmental Sciences (China). 2018;**63**:250-259

[27] Bhadra BN, Song JY, Lee SK, et al. Adsorptive removal of aromatic hydrocarbons from water over metal azolate framework-6-derived carbons. Journal of Hazardous Materials. 2018;**344**:1069-1077

[28] Zhang J, Yan X, Hu X, et al. Direct carbonization of Zn/Co zeolitic imidazolate frameworks for efficient adsorption of Rhodamine B. Chemical Engineering Journal. 2018;**347**:640-647

[29] Xu S, Lv Y, Zeng X, et al. ZIF-derived nitrogen-doped porous carbons as highly efficient adsorbents for removal of organic compounds from wastewater. Chemical Engineering Journal. 2017;**323**:502-511

[30] Jin L, Zhao X, Qian X, et al. Nickel nanoparticles encapsulated in porous carbon and carbon nanotube hybrids from bimetallic metal-organic-frameworks for highly efficient adsorption of dyes. Journal of Colloid and Interface Science. 2018;**509**:245-253

[31] Liu X, Wang C, Wang Z, et al. Nanoporous carbon derived from a metal organic framework as a new kind of adsorbent for dispersive solid phase extraction of benzoylurea insecticides. Microchimica Acta. 2015;**182**:1903-1910

[32] Lv Z, Wang H, Chen C, et al. Enhanced removal of uranium(VI) from aqueous solution by a novel Mg-MOF-74-derived porous MgO/carbon adsorbent. Journal of Colloid and Interface Science. 2019;**537**:A1-A10

[33] Ahsan MA, Deemer E, Fernandez-Delgado O, et al. Fe nanoparticles encapsulated in MOF-derived carbon for the reduction of 4-nitrophenol and methyl orange in water. Catalysis Communications. 2019;**130**:1-6. DOI: 10.1016/j.catcom.2019.105753

[34] Andrew Lin KY, Chen BC. Efficient elimination of caffeine from water using oxone activated by a magnetic and recyclable cobalt/carbon nanocomposite derived from ZIF-67. Dalton Transactions. 2016;**45**:3541-3551

[35] Bhadra BN, Lee JK, Cho C-W, et al. Remarkably efficient adsorbent for the removal of bisphenol A from water: Bio-MOF-1-derived porous carbon. Chemical Engineering Journal. 2018;**343**:225-234

[36] Torad NL, Hu M, Ishihara S, et al. Direct synthesis of MOF-derived nanoporous carbon with magnetic Co nanoparticles toward efficient water treatment. Small. 2014;**10**:2096-2107

[37] Chen D, Chen C, Shen W, et al. MOF-derived magnetic porous carbon-based sorbent: Synthesis, characterization, and adsorption behavior of organic micropollutants. Advanced Powder Technology. 2017;**28**:1769-1779

[38] Sarker M, Ahmed I, Jhung SH. Adsorptive removal of herbicides from water over nitrogen-doped carbon obtained from ionic liquid@ZIF-8. Chemical Engineering Journal. 2017;**323**:203-211

[39] Bhadra BN, Ahmed I, Kim S, et al. Adsorptive removal of ibuprofen and diclofenac from water using metal-organic framework-derived porous carbon. Chemical Engineering Journal. 2017;**314**:50-58

[40] Li S, Zhang X, Huang Y. Zeolitic imidazolate framework-8 derived nanoporous carbon as an effective and recyclable adsorbent for removal of ciprofloxacin antibiotics from water. Journal of Hazardous Materials. 2017;**321**:711-719

Chapter 8

Eco-Friendly Fluorescent Carbon Nanodots: Characteristics and Potential Applications

Adil Shafi, Sayfa Bano, Suhail Sabir, Mohammad Zain Khan and Mohammed Muzibur Rahman

Abstract

Carbon nanodots are zero-dimensional tiny particles of carbon with outstanding characteristics and potential applications. Carbon nanodots are fluorescent materials and possess unique characteristics such as biocompatibility, photostability, low toxicity, sustainable, and eco-friendly. Fluorescent carbon nanodots are emerging nanomaterials that show promising potential in bioimaging, optical sensing, information encryption and storage, photocatalysis, lasers, drug delivery, energy conversion, and photovoltaic applications. Carbon nanodots can be synthesized at very low cost through various sustainable approaches that employ inexpensive renewable resources as starting materials. Carbon nanodots are fascinating carbon-based materials that have received mass attention from past few years for their substantial applications in diverse fields. Carbon nanodots have a huge impact on both health and environmental applications because of their potential to serve as nontoxic replacements to traditional heavy metal-based quantum dots. Herein we highlight the intriguing characteristics and potential applications of fluorescent carbon nanodots in various fields and their perspective in future.

Keywords: carbon nanodots, biocompatible: nanomaterials, health, sustainable

1. Introduction

Carbon nanodots are the emerging class of carbon-based nanomaterials with small size and fascinating properties. Carbon nanodots are surface functionalized carbonaceous nanomaterials with tunable fluorescence and remarkable features [1]. Carbon nanodots are nowadays considered as the rising star in the nanocarbon family due to their sustainable, compatible, cost-effective, and benign nature [2]. Carbon was considered as black material unable to emit light up to the discovery of fluorescent carbon nanodots. The strong tunable luminescence combined with other fascinating properties has attracted considerable attention toward the fluorescent carbon nanodots [3]. Carbon nanodots also called graphene quantum dots share similar properties with graphene oxide with difference in size and shape, as the former is quasispherical in shape with nano-diameters [4]. The promising features of fluorescent carbon nanodots are remarkably assessed in various fields, namely, bioimaging, photocatalysis, optoelectronics, photovoltaics, drug delivery, and sensors [5].

Carbon nanodots were first discovered as a by-product in the purification of single-walled carbon nanotubes [6]. Since then, there have been a considerable increase in the interest of researchers to fabricate carbon-based fluorescent quantum dots, which can substitute the toxic and unstable metallic-based quantum dots (CdS, CdSe, CdTe, PbS, etc.). The fluorescent carbon nanodots possess many advantages over classical quantum dots in terms of biocompatibility, stability, solubility, inertness, cytotoxicity, drug delivery, and synthesis [7, 8]. In addition to that, carbon nanodots do not show photobleaching or blinking effects. In spite of that, they exhibit enhanced quantum yields and strong absorption in the UV-visible region. Although fluorescent quantum dots are prepared from different precursors and synthetic methodologies, the actual mechanism of fluorescence is still in debate [9]. Researchers have reported four possible mechanisms of fluorescence: quantum confinement effect or π domains; the surface state, determined by hybridization of carbon skeleton; the molecular state, determined by surface functionalities; and the crosslink-enhanced emission (CEE) effect [10]. Therefore, the strong blue-green and excitation dependent fluorescence depends on the synthetic method, experimental protocol, and the surface passivation of the carbon nanodot.

From past few years, extensive research is going on fluorescent carbon nanodots because of their intriguing properties and structural functionalities. On the surface of the carbon nanodot, several carboxyl, carbonyl, ammine, and amide moieties are present, which impart excellent solubility and biocompatibility [11]. The surfaces of the carbon nanodots can be modified and passivated with several organic, inorganic, polymer, or biological moieties, which in turn enhance their luminescent, sensing, and other properties. The carbon nanodot doped with a suitable heteroatom shows improved efficiency and enhanced radiative emission due to the shifting of Fermi level [12]. Based on specific surface morphology, carbon nanodots can be hydrophilic or hydrophobic. The hydrophilic nature of carbon nanodots makes them a promising material in diverse fields, thereby attracting the attention of researchers [13].

Among many fascinating applications in diverse fields, photoluminescence is the most researched and at the same time most debated application of these carbon nanodots. Fluorescent carbon nanodots exhibit strong and tunable fluorescence over a wide range of electromagnetic spectrum [14, 15]. Carbon nanodot fluorescence is sensitive to environmental conditions, solvents, temperature, pH, and external agents [16, 17]. Fluorescence is retarded by the agglomeration of particles and enhanced by the dispersion of particles [18]. The tunability of fluorescence provides multicolored blue-green emission, which spans entire visible region and is characterized by increased quantum yield [19]. The fluorescence shown by carbon nanodots can be efficiently quenched by oxidizing or reducing agents in solutions, indicating the electron acceptor or donor capability of carbon nanodots [20]. The redox property of these carbon nanodots can be exploited in light energy conversions, optoelectronics, photovoltaic devices, and in many related applications [21].

Fluorescent carbon dots have been synthesized by different conventional methodologies such as laser ablation, hot injection, hydrothermal, electrochemical, and acidic oxidation methods [22–26]. It is well known fact that carbon nanodots synthesized by different synthetic protocols using different precursors or modifications possess different physiochemical properties, which indicate their complex behavior. However, most methods face some limitation from environmental perspectives in using carbon precursors, synthetic procedure, and purification techniques [27, 28]. At present, extensive research is focused on using natural products to prepare fluorescent carbon dots. Preparation of fluorescent carbon dots by using renewable natural resources is cost-effective and can help in sustainable development of environment [29, 30]. The eco-friendly synthetic route combined with cost-effective approach can make the synthesized nanodots promising candidates for environmental remediation.

Carbon nanodots with a wide range of possible structures and architectures have been reported. The architecture of both the carbon core and surface functionalities plays a very important role in controlling the activity of the carbon nanodots [31]. The morphology of the carbon dots is nearly quasispherical, but the structure can be graphitic, turbostratic, amorphous, or crystalline. The hybridization of carbon atom may be sp^2 (turbostratic or graphitic carbon) or sp^3 (diamond-like carbon) [32]. In this chapter, the eco-friendly green synthetic approach along with fascinating characteristics and remarkable applications of fluorescent carbon dots is discussed. The surface functionalization and passivation to confer desirable properties to carbon nanodots have been highlighted. The future perspectives and the possible challenges, which can improve the physiochemical properties of carbon dots, have also been discussed.

2. Synthesis of fluorescent carbon nanodots

2.1 Conventional synthesis

Generally, the synthetic procedure of carbon nanodots is quite a tedious process, which involves several steps, surface passivation, and post-synthetic modifications [33, 34]. Also, the physiochemical properties and potential applications of carbon dots solely depend on the synthetic procedures [35]. Two conventional approaches have been employed to synthesize carbon nanodots, top-down approach, and bottom-up approach [36].

Top-down approach involves the cleavage of larger carbon cluster into smaller carbon fragments resulting in the formation of carbon nanodots with diameter less than 10 nm. The large molecules are fragmented by laser ablation, arc discharge, acid oxidation, or exfoliation methods (electrochemical, ultrasonic, solvothermal, and hydrothermal exfoliation methods) [2]. Contrarily bottom-up approach is based on several chemical reactions, which results in the conversion of small carbon precursors into nanoscale carbon dots. In bottom-down approach, the carbon nanodots are synthesized through partial dehydration and dehydrogenation by using microwave, solvothermal, pyrolysis, or thermal decomposition method [37, 38].

During preparation of carbon nanodots, several problems that need to be focused on control of size, proper functionalization of surface, and carbonaceous aggregation should be avoided [39, 40]. Size uniformity is important for uniform properties and mechanistic pathways, whereas surface functionalization is critical for solubility and surface applications. The size or surface properties can be optimized by post-treatment method, such as dialysis, gel-electrophoresis, ultracentrifugation, and filtration. Carbonaceous aggregation can be avoided by adopting electrochemical methods or confined pyrolysis methods [41] (**Table 1**).

2.2 Green synthesis

The conventional synthesis of carbon nanodots is uneconomical and unjustified because of high cost, tedious experimental protocols, post-preparation modifications, and time-consuming methodologies [45]. Therefore, it is imperative to design facile, eco-friendly, and sustainable methods to synthesize carbon nanodots. Green approach is one of the sustainable methods, which is very easy, cost-effective, eco-friendly, and nonlaborious [46]. From past few years, carbon nanodots were extensively synthesized using natural precursors through greener approach [47].

Synthetic method	Characteristics	Demerits	Literature
Laser ablation	• Top-down process • Fast and effective • Carbon NDs with strong fluorescence emissions • Tunable surface states	Low quantum yield, post-treatment is required	[42]
Chemical oxidation	• Top-down process • Most accessible • Various sources	Require drastic conditions, poor control over size, tedious protocol	[15]
Microwave irradiation	• Eco-friendly • Cost-effective and scalable	Modification not needed but poor control over size	[41]
Solvothermal/hydrothermal treatment	• Nontoxic • Cost-effective and eco-friendly	Poor control over particle size	[43]
Electrochemical carbonization	• Size control • Stable method • One-step method	Small molecule precursors, poor quantum yield	[44]

Table 1.
Characteristics of different synthetic methods used for preparation of carbon NDs.

Natural precursors include apple juice, orange juice, shrimps, sweet potatoes, garlic, aloe vera, and honey, which have been utilized as carbon precursors in the preparation of carbon nanodots [48]. In addition to this, plant wastes such as pomelo peel, willow bark, watermelon bark, and waste carbon paper have also been used as carbon precursors for synthesizing carbon nanodots. Biomaterials such as carbohydrates (starch, glucose, and sucrose) have also been utilized for the preparation of carbon nanodots [37]. The replacement of toxic and costly precursors by greener and natural precursors was considered to be the promising method for the synthesis of eco-friendly and potential florescent carbon dots.

2.3 Post-synthetic modifications

The synthesized carbon nanodots can be modified further to obtain nanodots with desirable properties. Generally, two strategies have been applied for post-synthetic modifications: suitable heteroatom doping and surface functionalization or modification.

Heteroatoms such as nitrogen, selenium, and sulfur can be incorporated into the nanodots by using suitable heteroatom-containing precursors [49, 50]. Carbon nanodots have also been co-doped with more than one type of heteroatoms through hydrothermal treatment [51]. Although heteroatom doping is facile method of structure modification, the actual mechanism is still unclear and results in irrational structural design.

Carbon nanodots can be selectively modified with the process of surface modification, which is easily controllable. The surface of carbon nanodots can be functionalized with several reactive intermediates through specific or nonspecific interactions [52]. The functionalization of carbon nanodots results in the tailoring of several properties, which in turn proves beneficial for several potential applications. Carbon nanodots have been passivated with oligomeric polymers to improve fluorescence emissions [53].

3. Morphology and composition

The morphology and structure of the carbon nanodots are exclusively dependent on reaction conditions of synthetic procedures [54]. These nanodots belong to the carbon nano family, existing in several possible substructures. Generally, carbon nanodots are considered to be quasispherical or spherical particles of carbon, which are less than 10 nm in size. The inner core of the carbon nanodots is crystalline or amorphous, but the surface layer covering the core contains several to many functional groups ranging from small amino groups to large fatty acid chains [32]. The crystalline structure of inner core was confirmed by several techniques such as scanning electron microscopy (SEM), transmission electron microscopy (TEM), X-ray diffraction (XRD), and amorphous nature, which were deduced by HRTEM technique [55]. The inner core is characterized by sp^2 (turbostratic or graphitic carbon) or sp^3 (diamond-like carbon). It has been reported that the inner core can acquire different possible structures at different conditions, particularly at high level of nitrogen doping [56].

The chemical composition of carbon nanodots also varies from one synthetic method to another. The composition of purified carbon nanodots has been reported as 36% carbon, 5.9% hydrogen, 9.4% nitrogen, and 44.7% oxygen. The oxygen content was significantly higher than the carbon nanodot synthesized from raw candle soot (91.7% C, 4.4% O, 1.8% H, and 1.8%N) [57] (**Figure 1**).

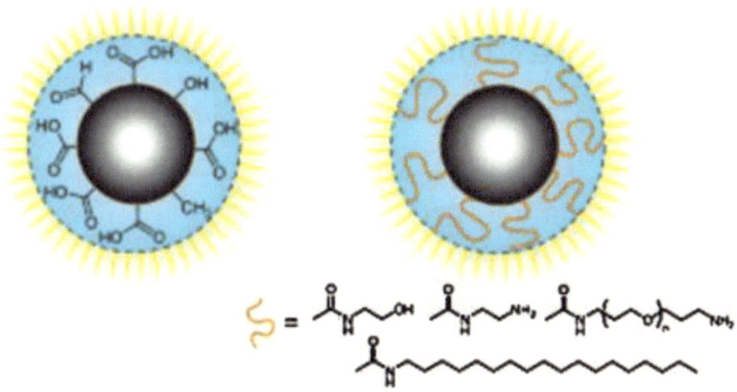

Figure 1.
Morphology and structure of carbon nanodot, taken from Ref. [58].

4. Characteristics of fluorescent carbon nanodots

Fluorescent carbon nanodots are tiny particles with fascinating characteristics and commendable properties. These nanodots are considered as versatile carbon members, which show a plethora of applications. In comparison with other semiconductor quantum dots, carbon nanodots show excellent properties, such as [58, 59]

- extraordinary optical, electrochemical, and electronic properties
- tunable fluorescence emission
- less toxicity
- excellent biocompatibility
- chemical robustness

- peculiar physiochemical properties
- solubility in water
- multiphoton excitation property
- eco-friendly nature

Carbon nanodots show maximum absorption in UV region with a tail extending to visible region. The optical absorption of carbon nanodots can be shifted to longer wavelength by surface passivation and doping with suitable dopants [60]. It has been observed that the nanodots exhibit size-dependent optical properties ranging from 200 to 500 cm^{-1}, when size is increased from 10 to 21 nm. Moreover, the reactive groups on surface also play a role in increasing the absorption properties of carbon nanodots [60, 61].

Fluorescent carbon nanodots exhibit remarkable wavelength-dependent photoluminescence properties that are totally different from Au nanodots, Ag nanodots, and other metallic nanodots. The fluorescence properties can be tuned through defect states without comprising with core structure. Functionalization of surface removes nonradiative redox recombination centers and increases the quantum yield [62]. Carbon nanodots show multiphoton excitation process, which results in emission of shorter wavelength light than the excitation wavelength. This up-conversion process is a sequential absorption process of one or more photons in the range of 320–425 nm. The photoluminescence from carbon nanodots is possible only when there is quantum confinement of surface energy traps, which becomes emissive upon stabilization through surface passivation [53].

The hydrophilicity of carbon nanodot is because of oxygen-containing functional groups over the surface, which imparts good solubility in water. Hydrophilicity modifies so many properties of carbon nanodots and makes them efficient fluorescent probes for several organic applications [63]. Recently, it has been reported the hydrophilic carbon nanodots can be converted into hydrophobic nanodots by covalent attachment of nonpolar solvents. Hydrophobic carbon nanodots have been used as efficient catalysts in organic synthesis [64].

Carbon nanodots possess remarkable redox properties, which make them efficient photocatalysts for degradation of organic pollutants, oxygen evolution, and CO_2 reduction [65]. The photocatalytic properties are enhanced by heteroatom doping, tuning of bandgap, and interfacial interactions. Carbon nanodots act as a photosensitizer for capturing solar light, thereby facilitating electron-hole separation [66]. Carbon nanodots have shown promising potential in water splitting due to the synergistic effects of several attributes. Zhang et al. introduced the metal-based semiconductor in photocatalysis as carbon dots decorated graphitic carbon nitride photocatalyst for the purification of water by phonol degradation. Muthulingam et al. described about highly efficient degradation of dyes by carbon quantum dots/N-doped zinc oxide photocatalyst and its compatibility on three different commercial dyes under daylight. Sharma et al. introduced about microwave-assisted fabrication of La/Cu/Zr/carbon dots trimetallic nanocomposites with their adsorption against photocatalytic efficiency for remediation of persistent organic pollutants.

4.1 Energy band structure

The energy band structure of carbon nanodots has been proposed on the basis of several computational and theoretical studies. The scheme of energy band structure of inner carbon core and the surface states has been depicted in **Figure 2**. It is

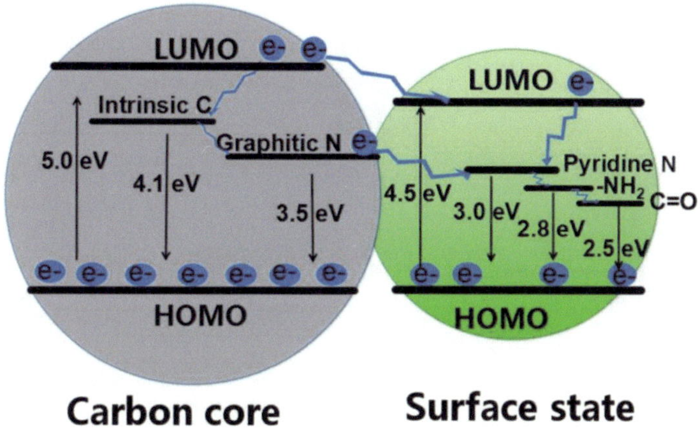

Figure 2.
Scheme depicting energy band structure of carbon nanodots, taken from Ref. [67].

evident from the figure that carbon nanodots exhibit five emission bands, which have been attributed to electron transitions at intrinsic carbon (305 nm), graphitic nitrogen (355 nm), pyridine nitrogen (410 nm), amino nitrogen (455 nm), and carboxyl carbons (500 nm) [67].

5. Applications

Fluorescent carbon nanodots have emerged as versatile carbon nanostructures with a wide range of potential applications. Based on their intriguing and fascinating properties such as biocompatibility, water solubility, and high stability, they are utilized as favorable materials in diverse fields. Carbon nanodots are promising materials and find substantial applications in bioimaging, photocatalysis, sensors (biosensors and chemical sensors), drug delivery, energy conversions, supercapacitors, LEDs, and many related processes [32, 68, 69]. In addition to this, carbon nanodots have shown great achievements in the field of food science in terms of food safety, nutrient management, and food toxicity [70–81]. The surface functionalizations with suitable reactive moieties have rendered carbon nanodots efficient in several biomedical applications such as in vivo and in vitro fluorescent probes and biomarkers. The applicability of carbon nanodots in biological and chemical sensing shows excellent results with respect to sensitivity, selectivity, stability, reproducibility, and response time.

5.1 Sensing

Carbon nanodots have been utilized as novel, efficient, and environment friendly fluorescent probes for the detection of trace quantities of chemical and biological analytes. Due to their fascinating and useful properties, carbon nanodots have been employed as biosensors for monitoring of glucose, DNA, phosphate, potassium, nitrite, and cellular copper with high selectivity and sensitivity [77, 82, 83]. The photoluminescence properties of carbon nanodots were investigated for the detection of various solvents (VOCs) [84]. It was reported that cyclic voltammetry technique was employed for selective and sensitive detection of glucose by using nitrogen-doped carbon nanodots with a LOD of 1–12 mM [85]. Boron-doped carbon nanodots were effectively utilized as chemosensors for trace detection of hydrogen

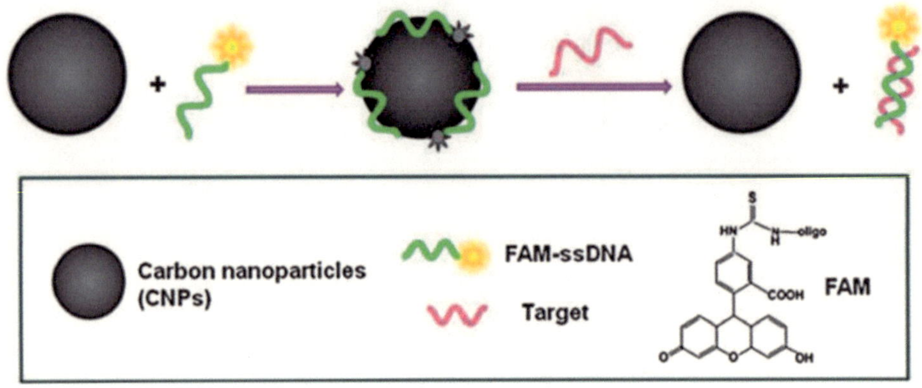

Figure 3.
Detection of nucleic acids by fluorescent carbon nanodots, taken from Ref. [90].

peroxide and glucose with very low detection limit [86]. Moreover, metal-doped carbon nanodots were effectively used as fluorescent sensors for sensing of dopamine, amoxicillin, catechol, pyridine, formaldehyde, pyrene, and so on [87, 88]. Metal-doped carbon nanodots have shown a considerable role in pH and temperature sensing in aqueous systems. Metal ions like Fe^{3+} are very important for the metabolism of living beings, and any fluctuation in its routine can be disastrous for human beings. Carbon nanodots can help in the detection of fluctuation of Fe^{3+} and thereby help in maintaining stable iron metabolism in the body [89].

Carbon nanodots can be used as a fluorescent nanosensor for nucleic acid detection with single-base mismatch [90–93]. The sensing is based on the adsorption of fluorescent labeled single-stranded DNA over carbon nanodots followed by fluorescence quenching and subsequent hybridization with its target to form double-stranded DNA (**Figure 3**).

5.2 Photocatalysis

Due to efficient redox properties, carbon nanodots have been efficiently employed as photocatalysts for harnessing solar energy in organic pollutant degradation. Carbon nanodots upon irradiation generate electron hole pairs, which can be subsequently utilized for multiple applications in pollutant degradation, CO_2 reduction, and photo catalytic water splitting [94, 95]. Carbon nanodots have been considered as excellent photocatalysts with a strong absorption in the wide range of electromagnetic spectrum. However, due to poor electron transfer inside the carbon nanodots, the application has been impeded. In order to increase the efficiency of carbon nanodots and to make them better photocatalysts, their electronic structure is modified by adopting several strategies namely, metal ion doping, heterostructure formation, composite formation, and so on [8, 95]. Doped carbon nanodots show efficient electronic properties with a strong visible light absorption and show low recombination of charge carriers [96]. Nitrogen-doped carbon nanodots in comparison with bare carbon nanodots show efficient visible light photo catalytic degradation of methyl orange. It has been also reported that carbon nanodots in the size range of 1–4 nm showed good photocatalytic oxidation of benzyl alcohol to benzaldehyde in the presence of H_2O_2 [97]. The conversion efficiency under NIR light was observed to be 92–100%, confirming better redox properties of carbon nanodots. The proposed mechanism for the conversion has been demonstrated in **Figure 4**.

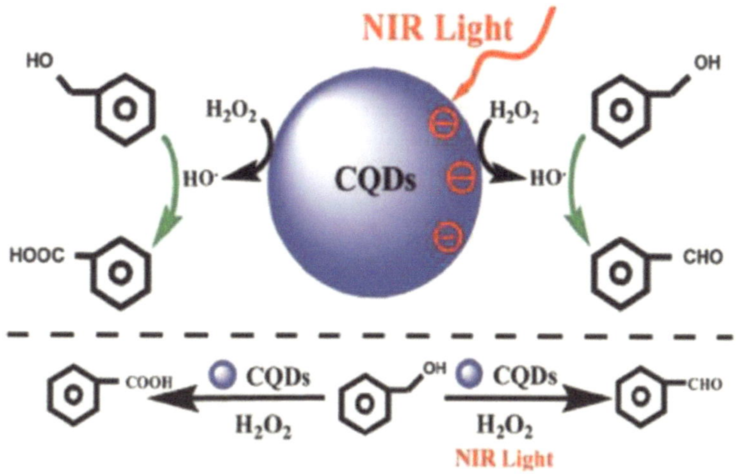

Figure 4.
Carbon nanodot supported oxidation of benzyl alcohol to benzaldehyde in the presence of NIR light, taken from Ref. [97].

5.3 Optronics

White light emitting diode can help in saving a lot of energy, but conventional diodes with rare earth metals suffer a drawback in terms of cost, stability, and toxicity. Because of low cost, eco-friendliness, high-quantum yield and low toxicity, carbon nanodots are replacing the traditional white light emitting diodes. Carbon nanodots are the promising materials to replace phosphors in white light emitting diodes with toxic elements such as cadmium and lead [98]. Carbon nanodots serve as a potential candidate in dye-sensitized solar cells, supercapacitors, and organic solar cells [99, 100]. Carbon nanodots doped with nitrogen or coupled with polymer matrix show a considerable attention in LEDs because of flexibility, thermal stability, and robustness.

6. Conclusion

In summary, fluorescent carbon nanodots are the members of carbon family with fascinating and remarkable properties. Although several protocols have been discussed about their synthesis, the size control and precise morphologies have not been attained yet. It is noteworthy to mention that the green synthesis of carbon nanodots has proved facile and effective in controlling the size and properties. Fluorescent carbon nanodots are unique tiny materials with extraordinary characteristics and commendable properties. The properties of carbon are explored in several fields. Carbon nanodots have shown explicit potential in biomedical, photovoltaic, optoelectronic, and electrochemical fields. In addition, the excellent redox properties and light harnessing potentiality have rendered them potential candidates for photo catalytic applications. Furthermore, the newly discovered chroptical properties of carbon nanodots will certainly find promising applications in both biomedical and electronic fields. Despite of the peculiar and remarkable applications in diverse fields, several properties of the carbon nanodots are still unclear. In future, extensive studies are needed to elucidate the possible mysteries and novel applications of carbon nanodots.

Acknowledgements

The authors are highly thankful to the Department of Chemistry, Aligarh Muslim University for providing research facilities. The authors also acknowledge the authors whose work in the form of images has been reproduced in this chapter.

Author details

Adil Shafi[1], Sayfa Bano[1], Suhail Sabir[1], Mohammad Zain Khan[1] and Mohammed Muzibur Rahman[2]*

1 Environmental Research Laboratory, Department of Chemistry, Aligarh Muslim University, Aligarh, India

2 Department of Chemistry, Faculty of Science, King Abdulaziz University, Jeddah, Saudi Arabia

*Address all correspondence to: mmrahman@kau.edu.sa; mmrahmanh@gmail.com

References

[1] Baker SN, Baker GA. Luminescent carbon nanodots: Emergent nanolights. Angewandte Chemie, International Edition. 2010;**49**(38):6726-6744

[2] Li H, Kang Z, Liu Y, Lee S-T. Carbon nanodots: Synthesis, properties and applications. Journal of Materials Chemistry. 2012;**22**(46):24230-24253

[3] Sciortino A, Cannizzo A, Messina F. Carbon nanodots: A review—From the current understanding of the fundamental photophysics to the full control of the optical response. C. 2018;**4**(4):67

[4] d'Amora M, Giordani S. Carbon nanomaterials for nanomedicine. In: Smart Nanoparticles for Biomedicine. Elsevier; 2018. pp. 103-113

[5] Karfa P, De S, Majhi KC, Madhuri R, Sharma PK. Functionalization of carbon nanostructures. In: Comprehensive Nanoscience and Nanotechnology. Elsevier; 2019. pp. 123-144

[6] Sagbas S, Sahiner N. Carbon dots: Preparation, properties, and application. In: Nanocarbon and its Composites. Elsevier; 2019. pp. 651-676

[7] Papaioannou N et al. Structure and solvents effects on the optical properties of sugar-derived carbon nanodots. Scientific Reports. 2018;**8**(1):6559

[8] Sha A, Ahmad N, Sultana S, Sabir S, Khan MZ. Ag 2 S - sensitized NiO–ZnO heterostructures with enhanced visible light photocatalytic activity and acetone sensing property. ACS omega. 2019;**4**(7):12905-12918

[9] Sun X, Lei Y. Fluorescent carbon dots and their sensing applications. TrAC, Trends in Analytical Chemistry. 2017;**89**:163-180

[10] Zhu S, Song Y, Zhao X, Shao J, Zhang J, Yang B. The photoluminescence mechanism in carbon dots (graphene quantum dots, carbon nanodots, and polymer dots): Current state and future perspective. Nano Research. 2015;**8**(2):355-381

[11] Shi L et al. Green-fluorescent nitrogen-doped carbon nanodots for biological imaging and paper-based sensing. Analytical Methods. 2017;**9**(14):2197-2204

[12] Xu Y et al. Nitrogen-doped carbon dots: A facile and general preparation method, photoluminescence investigation, and imaging applications. Chemistry—A European Journal. 2013;**19**(7):2276-2283

[13] Panniello A et al. Luminescent oil-soluble carbon dots toward white light emission: A spectroscopic study. Journal of Physical Chemistry C. 2018;**122**(1):839-849

[14] Liu C, Zhang P, Tian F, Li W, Li F, Liu W. One-step synthesis of surface passivated carbon nanodots by microwave assisted pyrolysis for enhanced multicolor photoluminescence and bioimaging. Journal of Materials Chemistry. 2011;**21**(35):13163-13167

[15] Qiao ZA et al. Commercially activated carbon as the source for producing multicolor photoluminescent carbon dots by chemical oxidation. Chemical Communications. 2010;**46**(46):8812-8814

[16] Jiang P, Tian ZQ, Zhu CN, Zhang ZL, Pang DW. Emission-tunable near-infrared Ag 2S quantum dots. Chemistry of Materials. 2012;**24**(1):3-5

[17] Pan D, Zhang J, Li Z, Wu C, Yan X, Wu M. Observation of pH-, solvent-, spin-, and excitation-dependent blue photoluminescence from carbon nanoparticles.

Chemical Communications. 2010;**46**(21):3681-3683

[18] Bourlinos AB, Zbořil R, Petr J, Bakandritsos A, Krysmann M, Giannelis EP. Luminescent surface quaternized carbon dots. Chemistry of Materials. 2012;**24**(1):6-8

[19] Pan L et al. Truly fluorescent excitation-dependent carbon dots and their applications in multicolor cellular imaging and multidimensional sensing. Advanced Materials. 2015;**27**(47):7782-7787

[20] Wang X et al. Photoinduced electron transfers with carbon dots. Chemical Communications. 2009;**46**(25):3774-3776

[21] Li Y et al. An electrochemical avenue to green-luminescent graphene quantum dots as potential electron-acceptors for photovoltaics. Advanced Materials. 2011;**23**(6):776-780

[22] Liu C et al. Strong infrared laser ablation produces white-light-emitting materials via the formation of silicon and carbon dots in silica nanoparticles. Journal of Physical Chemistry C. 2015;**119**(15):8266-8272

[23] He D et al. Dielectric barrier discharge-assisted one-pot synthesis of carbon quantum dots as fluorescent probes for selective and sensitive detection of hydrogen peroxide and glucose. Talanta. 2015;**142**:51-56

[24] Mehta VN, Jha S, Kailasa SK. One-pot green synthesis of carbon dots by using Saccharum officinarum juice for fluorescent imaging of bacteria (*Escherichia coli*) and yeast (*Saccharomyces cerevisiae*) cells. Materials Science and Engineering: C. 2014;**38**(1):20-27

[25] Xu L, Yin H, Ma W, Wang L, Kuang H, Xu C. MRI biosensor for lead detection based on the DNAzyme-induced catalytic reaction. The Journal of Physical Chemistry. B. 2013;**117**(46):14367-14371

[26] Canevari TC, Nakamura M, Cincotto FH, De Melo FM, Toma HE. High performance electrochemical sensors for dopamine and epinephrine using nanocrystalline carbon quantum dots obtained under controlled chronoamperometric conditions. Electrochimica Acta. 2016;**209**:464-470

[27] Wu M et al. Preparation of functionalized water-soluble photoluminescent carbon quantum dots from petroleum coke. Carbon. 2014;**78**:480-489

[28] Gedda G, Lee CY, Lin YC, Wu HF. Green synthesis of carbon dots from prawn shells for highly selective and sensitive detection of copper ions. Sensors and Actuators B: Chemical. 2016;**224**:396-403

[29] Chen Y, Wu Y, Weng B, Wang B, Li C. Facile synthesis of nitrogen and sulfur co-doped carbon dots and application for Fe(III) ions detection and cell imaging. Sensors and Actuators B: Chemical. 2016;**223**:689-696

[30] Aslandaş AM, Balci N, Arik M, Şakiroılu H, Onganer Y, Meral K. Liquid nitrogen-assisted synthesis of fluorescent carbon dots from blueberry and their performance in Fe 3+ detection. Applied Surface Science. 2015;**356**:747-752

[31] Yao B, Huang H, Liu Y, Kang Z. Carbon dots: A small conundrum. Trends in Chemistry. 2019;**1**(2):10

[32] Lim SY, Shen W, Gao Z. Carbon quantum dots and their applications. Chemical Society Reviews. 2015;**44**(1):362-381

[33] Zhu S et al. Surface chemistry routes to modulate the photoluminescence of graphene quantum dots: From

fluorescence mechanism to up-conversion bioimaging applications. Advanced Functional Materials. 2012;**22**(22):4732-4740

[34] Huang H et al. Histidine-derived nontoxic nitrogen-doped carbon dots for sensing and bioimaging applications. Langmuir. 2014;**30**(45):13542-13548

[35] Vasimalai N et al. Green synthesis of fluorescent carbon dots from spices for in vitro imaging and tumour cell growth inhibition. Beilstein Journal of Nanotechnology. 2018;**9**(1):530-544

[36] Wang Y, Hu A. Carbon quantum dots: Synthesis, properties and applications. Journal of Materials Chemistry C. 2014;**2**:6921

[37] Peng H, Travas-Sejdic J. Simple aqueous solution route to luminescent carbogenic dots from carbohydrates. Chemistry of Materials. 2009;**21**(23): 5563-5565

[38] Lecroy GE et al. Toward structurally defined carbon dots as ultracompact fluorescent probes. ACS Nano. 2014;**8**(5):4522-4529

[39] Wang H, Gao Q. Synthesis, characterization and energy-related applications of carbide-derived carbons obtained by the chlorination of boron carbide. Carbon. 2009;**47**(3):820-828

[40] Zhou J et al. Tailoring multi-wall carbon nanotubes for smaller nanostructures. Carbon. 2009;**47**(3):829-838

[41] Zhu H, Wang X, Li Y, Wang Z, Yang F, Yang X. Microwave synthesis of fluorescent carbon nanoparticles with electrochemiluminescence properties. Chemical Communications. 2009;**4**(34):5118-5120

[42] Hu SL, Niu KY, Sun J, Yang J, Zhao NQ, Du XW. One-step synthesis of fluorescent carbon nanoparticles by laser irradiation. Journal of Materials Chemistry. 2009;**19**(4):484-488

[43] Titirici MM, Antonietti M. Chemistry and materials options of sustainable carbon materials made by hydrothermal carbonization. Chemical Society Reviews. 2010;**39**(1):103-116

[44] Zhou J et al. An electrochemical avenue to blue luminescent nanocrystals from multiwalled carbon nanotubes (MWCNTs). Journal of the American Chemical Society. 2007;**129**(4):744-745

[45] Devi S, Gupta RK, Paul AK, Tyagi S. Waste carbon paper derivatized carbon quantum dots/(3-Aminopropyl) triethoxysilane based fluorescent probe for trinitrotoluene detection. Materials Research Express. 2019;**6**(2):025605

[46] Vikneswaran R, Ramesh S, Yahya R. Green synthesized carbon nanodots as a fluorescent probe for selective and sensitive detection of iron(III) ions. Materials Letters. 2014;**136**:179-182

[47] Du W et al. Green synthesis of fluorescent carbon quantum dots and carbon spheres from pericarp. Science China. Chemistry. 2015;**58**(5):863-870

[48] Kumar A, Chowdhuri AR, Laha D, Mahto TK, Karmakar P, Sahu SK. Green synthesis of carbon dots from *Ocimum sanctum* for effective fluorescent sensing of Pb2+ ions and live cell imaging. Sensors and Actuators B: Chemical. 2017;**242**:679-686

[49] Ma Z, Ming H, Huang H, Liu Y, Kang Z. One-step ultrasonic synthesis of fluorescent N-doped carbon dots from glucose and their visible-light sensitive photocatalytic ability. New Journal of Chemistry. 2012;**36**(4):861-864

[50] Li F, Li T, Sun C, Xia J, Jiao Y, Xu H. Selenium-doped carbon quantum dots for free-radical scavenging. Angewandte Chemie, International Edition. 2017;**56**(33):9910-9914

[51] Li F et al. Highly fluorescent chiral N-S-doped carbon dots from cysteine: Affecting cellular energy metabolism. Angewandte Chemie, International Edition. 2018;**57**(9):2377-2382

[52] Chua CK et al. Synthesis of strongly fluorescent graphene quantum dots by cage-opening buckminsterfullerene. ACS Nano. 2015;**9**(3):2548-2555

[53] Sun YP et al. Quantum-sized carbon dots for bright and colorful photoluminescence. Journal of the American Chemical Society. 2006;**128**(24):7756-7757

[54] Cayuela A, Soriano ML, Carrillo-Carrión C, Valcárcel M. Semiconductor and carbon-based fluorescent nanodots: The need for consistency. Chemical Communications. 2016;**52**(7):1311-1326

[55] Shinde DB, Pillai VK. Electrochemical preparation of luminescent graphene quantum dots from multiwalled carbon nanotubes. Chemistry—A European Journal. 2012;**18**(39):12522-12528

[56] Sciortino A et al. β-C3N4 nanocrystals: Carbon dots with extraordinary morphological, structural, and optical homogeneity. Chemistry of Materials. 2018;**30**(5):1695-1700

[57] Liu H, Ye T, Mao C. Fluorescent carbon nanoparticles derived from candle soot. Angewandte Chemie, International Edition. 2007;**46**(34):6473-6475

[58] Namdari P, Negahdari B, Eatemadi A. Synthesis, properties and biomedical applications of carbon-based quantum dots: An updated review. Biomedicine and Pharmacotherapy. 2017;**87**:209-222

[59] Xiao L, Sun H. Novel properties and applications of carbon nanodots. Nanoscale Horizons. 2018;**3**:565

[60] Peng J et al. Graphene quantum dots derived from carbon fibers. Nano Letters. 2012;**12**(2):844-849

[61] Riggs JE, Guo Z, Carroll DL, Sun YP. Strong luminescence of solubilized carbon nanotubes[2]. Journal of the American Chemical Society. 2000;**122**(24):5879-5880

[62] Loh KP, Bao Q, Eda G, Chhowalla M. Graphene oxide as a chemically tunable platform for optical applications. Nature Chemistry. 2010;**2**(12):1015-1024

[63] Zheng M et al. Self-targeting fluorescent carbon dots for diagnosis of brain cancer cells. ACS Nano. 2015;**9**(11):11455-11461

[64] Bourlinos AB, Stassinopoulos A, Anglos D, Zboril R, Georgakilas V, Giannelis EP. Photoluminescent carbogenic dots. Chemistry of Materials. 2008;**20**(14):4539-4541

[65] Jeon SJ et al. Modulating the photocatalytic activity of graphene quantum dots via atomic tailoring for highly enhanced photocatalysis under visible light. Advanced Functional Materials. 2016;**26**(45):8211-8219

[66] Ong WJ et al. Unravelling charge carrier dynamics in protonated g-C3N4 interfaced with carbon nanodots as co-catalysts toward enhanced photocatalytic CO2 reduction: A combined experimental and first-principles DFT study. Nano Research. 2017;**10**(5):1673-1696

[67] Yu J et al. Luminescence mechanism of carbon dots by tailoring functional groups for sensing Fe3+ ions. Nanomaterials. 2018;**8**(4):233

[68] Du Y, Guo S. Chemically doped fluorescent carbon and graphene quantum dots for bioimaging, sensor, catalytic and photoelectronic applications. Nanoscale. 2016;**8**(5): 2532-2543

[69] Kailasa SK, Mehta VN, Hasan N, Wu HF. Applications of carbon dots in biosensing and cellular imaging. In: Nanobiomaterials in Medical Imaging: Applications of Nanobiomaterials. Elsevier Inc; 2016. pp. 339-364

[70] Qu JH, Wei Q, Sun DW. Carbon dots: Principles and their applications in food quality and safety detection. Critical Reviews in Food Science and Nutrition. 2018;**58**(14):2466-2475

[71] Gao X, Yang L, Petros JA, Marshall FF, Simons JW, Nie S. In vivo molecular and cellular imaging with quantum dots. Current Opinion in Biotechnology. 2005;**16**(1):63-72

[72] Hardman R. A toxicologic review of quantum dots: Toxicity depends on physicochemical and environmental factors. Environmental Health Perspectives. 2006;**114**(2):165-172

[73] Ghosal K, Ghosh A. Carbon dots: The next generation platform for biomedical applications. Materials Science and Engineering C. 2019;**96**:887-903

[74] Boakye-Yiadom KO et al. Carbon dots: Applications in bioimaging and theranostics. International Journal of Pharmaceutics. 2019;**564**:308-317

[75] Tuerhong M, XU Y, YIN XB. Review on carbon dots and their applications. Chinese Journal of Analytical Chemistry. 2017;**45**(1):139-150

[76] Hsu PC, Shih ZY, Lee CH, Chang HT. Synthesis and analytical applications of photoluminescent carbon nanodots. Green Chemistry. 2012;**14**(4):917-920

[77] Yang Z et al. Controllable synthesis of fluorescent carbon dots and their detection application as nanoprobes. Nano-Micro Letters. 2013;**5**(4):247-259

[78] Rosenthal SJ, Chang JC, Kovtun O, McBride JR, Tomlinson ID. Biocompatible quantum dots for biological applications. Chemistry & Biology. 2011;**18**(1):10-24

[79] Xu Y, Li YH, Wang Y, Cui JL, Zhang YK. 13 C-engineered carbon quantum dots for in vivo magnetic resonance and fluorescence dual-respons. Analyst. 2014;**139**(20):5134-5139

[80] Liu JJ, Li D, Zhang K, Yang M, Sun H, Yang B. One-step hydrothermal synthesis of nitrogen-doped conjugated carbonized polymer dots with 31% efficient red emission for in vivo imaging. Small. 2018;**14**(15):1703919

[81] Huang X et al. Effect of injection routes on the biodistribution, clearance, and tumor uptake of carbon dots. ACS Nano. 2013;7(7):5684-5693

[82] Zhou L, Lin Y, Huang Z, Ren J, Qu X. Carbon nanodots as fluorescence probes for rapid, sensitive, and label-free detection of Hg 2+ and biothiols in complex matrices. Chemical Communications. 2012;**48**(8):1147-1149

[83] Zhao J et al. Ultrafast spontaneous emission modulation of graphene quantum dots interacting with Ag nanoparticles in solution. Applied Physics Letters. 2016;**109**(2):021905

[84] Campos BB et al. Fluorescent chemosensor for pyridine based on N-doped carbon dots. Journal of Colloid and Interface Science. 2015;**458**:209-216

[85] Ji H, Zhou F, Gu J, Shu C, Xi K, Jia X. Nitrogen-doped carbon dots as a new substrate for sensitive glucose determination. Sensors. 2016;**16**(5):630

[86] Shan X, Chai L, Ma J, Qian Z, Chen J, Feng H. B-doped carbon quantum dots as a sensitive fluorescence probe for hydrogen peroxide and glucose detection. Analyst. 2014;**139**(10):2322-2325

[87] Li H et al. Fluorescent N-doped carbon dots for both cellular imaging and highly-sensitive catechol detection. Carbon. 2015;**91**:66-75

[88] Jiang Y, Wang B, Meng F, Cheng Y, Zhu C. Microwave-assisted preparation of N-doped carbon dots as a biosensor for electrochemical dopamine detection. Journal of Colloid and Interface Science. 2015;**452**:199-202

[89] Ru GJ, Xin Q, Rui JX, Shah H. Single precursor-based luminescent nitrogen-doped carbon dots and their application for iron (III) sensing. Arabian Journal of Chemistry. 2019. DOI: https://doi.org/10.1016/j.arabjc.2019.06.004

[90] Li H, Zhang Y, Wang L, Tian J, Sun X. Nucleic acid detection using carbon nanoparticles as a fluorescent sensing platform. Chemical Communications. 2011;**47**(3):961-963

[91] Thakur M et al. Antibiotic conjugated fluorescent carbon dots as a theranostic agent for controlled drug release, bioimaging, and enhanced antimicrobial activity. Journal of Drug Delivery. 2014;**2014**:1-9

[92] Tang J et al. Carbon nanodots featuring efficient FRET for real-time monitoring of drug delivery and two-photon imaging. Advanced Materials. 2013;**25**(45):6569-6574

[93] Zheng M et al. Integrating oxaliplatin with highly luminescent carbon dots: An unprecedented theranostic agent for personalized medicine. Advanced Materials. 2014;**26**(21):3554-3560

[94] Liu J et al. Carbon quantum dot/silver nanoparticle/polyoxometalate composites as photocatalysts for overall water splitting in visible light. ChemCatChem. 2014;**6**(9):2634-2641

[95] Zhang Z, Zheng T, Li X, Xu J, Zeng H. Progress of carbon quantum dots in photocatalysis applications. Particle and Particle Systems Characterization. 2016;**33**(8):457-472

[96] Bourlinos AB, Stassinopoulos A, Anglos D, Zboril R, Karakassides M, Giannelis EP. Surface functionalized carbogenic quantum dots. Small. 2008;**4**(4):455-458

[97] Li H, Liu R, Lian S, Liu Y, Huang H, Kang Z. Near-infrared light controlled photocatalytic activity of carbon quantum dots for highly selective oxidation reaction. Nanoscale. 2013;**5**(8):3289-3297

[98] Zhang X et al. Color-switchable electroluminescence of carbon dot light-emitting diodes. ACS Nano. 2013;**7**(12):11234-11241

[99] Zhu Y et al. A carbon quantum dot decorated RuO 2 network: Outstanding supercapacitances under ultrafast charge and discharge. Energy & Environmental Science. 2013;**6**(12):3665-3675

[100] Ma Z et al. Bioinspired photoelectric conversion system based on carbon-quantum-dot- doped dye-semiconductor complex. ACS Applied Materials & Interfaces. 2013;**5**(11):5080-5084

Chapter 9

The Role of Mangroves Forests in Decarbonizing the Atmosphere

Charles Nyanga

Abstract

Mangrove forests occupy approximately not more than 1% of the world's forested land, according to experts. These important ecosystems are currently being lost at an alarming rate. Aquaculture, urban development, agriculture, and industrial development have been observed to be the major causes of these mangrove losses. Mangroves are an important source of ecosystem goods and services, among which are carbon sequestration, providing breeding and nursery grounds for several species of flora and fauna, materials, medicines, and climate change impact protection. Carbon dioxide capturing and sequestration is a system of man-made processes to reduce carbon dioxide emissions from utilities which use coal and gas. Mangroves can actually do this as natural carbon capture and sequestration (CCS) agents for mankind.

Keywords: mangroves, ecosystems, capture, sequestration

1. Introduction

The term mangrove (Rhizophoraceae) refers to various plant species (trees and shrubs) which are tolerant to salty waters and normally grow in the intertidal zones of coastlines belonging to tropical and subtropical sheltered coastlines [1]. The term is applied to both individual plants and entire ecosystems occupied by mangroves. The area covered by mangroves is referred to as the mangal. Mangroves cover less than 1% of tropical forests worldwide [1]. Mangroves are a well-adapted plant species that grow in fresh, brackish, and salty water wetlands. They occur dominantly in brackish and salty water wetlands since this tends to eliminate competition from plants and shrubs which are adapted to freshwater wetlands. The mangroves' adaptation mechanisms include salt coping mechanisms which enable them to filter out more than 90% of the salt in seawater, water hoarding mechanism which enables them to hoard water in thick succulent leaves, and breathing in a variety of ways using snorkel-like parts [2].

According to [3], the *Rhizophora* species are the red mangroves. The red mangroves develop better within brackish or marine water bodies, while the other species known as *Avicennia* are the black mangroves which develop better near the land side of the estuaries. Both of these species are found in areas which receive flooding and which are shallow during high tide. For a proper development they require average coldest monthly temperatures of more than 20° C and seasonal temperature variations not exceeding 5° C. In addition, they also require fine grained growing media and strong wave and tidal current-free shores. Salty water and large tidal range are also required for the development of mangroves.

2. Geographical distribution of mangroves

Mangrove forests are confined to regions 30° north and 30° south of the equator. The total area occupied by mangrove forests in the world is 18.1 million ha. South and Southeast Asia has a total of 7.5173 million ha of mangrove forest area. This is equal to 41.5% of the world's total mangrove forest area [4]. The region where mangroves are found is known to be sheltered and receives high amounts of rainfall. In terms of expanse, the largest areas of mangroves are in Asia, which tops the list in terms of size, Asia is followed by Africa, and the last on the list comes South America. At country level, there are four countries whose mangrove areas amount to 41% of the total area of mangroves in the world. These countries are Indonesia, Brazil, Nigeria, and Australia [5].

Mangroves have been restricted to tropics and subtropics because these special trees are very sensitive to the cold. They are predominantly found in two regions within the tropics and subtropics; these two regions are, namely, the Indo-West Pacific which has 58 species and the Atlantic-East Pacific which has 12 species [6].

3. Ecology of mangroves

True mangroves have been observed to occur in nature as 54–75 species. Mangroves which are healthy are very important to heathy marine ecology. This is so due to the detritus which originates from the fallen leaves and branches from mangroves. This detritus provides support to a lot of sea creatures [7].

Additionally, mangroves form valuable ecosystems. These ecosystems play many important roles apart from providing detritus. They are habitats to several species of plants and animals alike. The competition from other plants is reduced to almost zero. This means that there are very few plants which are able to survive in mangrove ecosystems. However, there are quite a number of animal species which are able to thrive in mangrove forests. The ecosystems therefore act as breeding grounds for birds, reptiles, crustaceans, fish, and insects. They are also home to the juveniles as well as entire communities of these creatures [7].

4. Biology of mangroves

The biology of mangroves is very interesting. They have various adaptations that enable them to survive. Of the 110 species which are known as mangroves, only 54–72 species in 10 genera are known to be true mangroves [8]. These 54–72 species belong to 16 families as well, and they seldom occur outside mangrove habitat [5].

4.1 Adaptation for support in soft waterlogged sediments

Mangroves have a wide variety of adaptations, one of which is the dense root system which gives them ability to stand upright in soft, waterlogged coastal sediment [5]. The sediment is so fine and soft. This is also another way of excluding other plants from being able to develop where they grow, and therefore this eliminates competition for the rarely available nutrients and oxygen (**Figure 1**).

4.2 Adaptation for reproduction and survival of offspring

The mangroves' adaptation in terms of reproduction appears in several forms. Mangroves are viviparous. The occurrence of viviparity in mangroves means that the seeds break their dormancy while still on the parent plant. Another way reproduction and survival of offspring is ensured is seed dispersal by the flowing currents and varying tides of the water which tend to throw the germinating seeds far and wide [5].

The aforementioned adaptation ways including buoyant or floating seeds which are able to float horizontally and later switch to a vertical position securing anchorage in the fine and soft substrate for growth tend to ensure survival of the offspring (**Figure 2**).

4.3 Adaptation to low quantities of oxygen

The mangroves' various adaptations depend on the species. The red mangroves have lenticels which are propped above the water to absorb oxygen through the bark of the tree, while black mangroves have pneumatophores. The pneumatophores are root-like structures which are specialized and stick out of the saline water to enable

Figure 1.
The figure is showing the way mangroves are adapted to survive in the waterlogged soft substrate for growth by having widely spaced and dense root system. Photo credit: Beatrice Njeri Obegi, Research Scientist, Kenya Marine and Fisheries Research Institute.

Figure 2.
The figure is showing the way mangroves are adapted to increase offspring survival. In the foreground some offspring can be seen germinating on the parent and have quick growth rate. Photo credit: Beatrice Njeri Obegi, Research Scientist, Kenya Marine and Fisheries Research Institute.

breathing by the plant. They are also covered with lenticels. The pneumatophores can reach up to 30 cm in height. Mangroves have four types of breathing tubes, namely, stilt or prop type, snorkel or peg type, knee type, and ribbon type [8]. These are so well adapted as to enable mangroves to breath depending on the type of environmental survival requirements.

4.4 Adaptation for efficient nutrient uptake

One of the major requirements for plant growth is need for enough nutrients. Mangroves have developed processes which enable them to take up and utilize nutrients efficiently from the soil which have low nutrient content. They have also developed nutrient conserving processes which include evergreenness, reabsorption of nutrients, immobilization of nutrients, and high root/shoot ratio [9]. Mangroves are also opportunistic in utilizing nutrients (**Figure 3**).

Figure 3.
The figure is showing the way mangroves are adapted to survive by efficient nutrient uptake and use. This figure is clearly showing the large root/shoot ratio. Photo credit: Beatrice Njeri Obegi, Research Scientist, Kenya Marine and Fisheries Research Institute.

Figure 4.
The figure is showing the way mangroves are adapted to lessen internal water loss. The shiny and succulent leaves can be seen on the left side of the photo. Photo credit: Beatrice Njeri Obegi, Research Scientist, Kenya Marine and Fisheries Research Institute.

4.5 Adaptation to the saline environment

Mangroves have evolved many ways of adapting to saltwater. The major ways they have are that they filter out the salt from the water and take up only the almost pure water. Mangroves are able to filter out more than 90 percent of the salt in saline water as it enters the roots. The other way they have adapted to the saline condition is by excreting the salt from the leaves.

4.6 Adaptation to limit internal water loss

Mangroves hoard water in a variety of ways. They store water in their thick succulent leaves. The leaves of some mangrove species have a waxy coating to seal in water, and evaporation is reduced to a minimum. Yet some other species have tiny hairs to deflect wind and sunlight (**Figure 4**). In addition they also have openings on the underside of their leaves away from the sunlight [10].

5. Carbon capture and storage technologies: the state of the art and future prospects

Currently there are many technologies available to achieve the decarburization of the atmosphere to maintain the temperature of the atmosphere at a value or range of values which are not exceeding the 2°C threshold. According to [11], carbon capture and sequestration (CCS) is a set of technologies which have been articulated widely by several authors as having potential for reducing the concentrations of carbon dioxide in the atmosphere. The hardworking writers agree that these technologies have the potential of playing a pivotal role in helping mankind to meet the climate change targets. Additionally, [11] reports that carbon capture and sequestration (CCS) has the capability to deliver low carbon heat and power and can also help in decarbonizing industries (especially the steel and coal industries) with further applications to facilitate in reducing the net removal of CO_2 from the atmosphere.

The technologies of carbon capture and sequestration, according to [12], should include the storage of CO_2 for several hundreds of years for climate change mitigation to be meaningful to human beings. Further, [12] reports that carbon capture and storage involves three stages, namely, (i) capture of CO_2, (ii) transport of CO_2, and (iii) storage of CO_2 (or injection of CO_2). However, Herzbog [13] adds a fourth stage which is the stage of monitoring of the CO_2 stored.

5.1 Capture of CO_2

The capture of CO_2 is the first step in carbon capture and storage. This step according to [13, 14] is essentially to ensure that the CO_2 is transported and stored successfully and economically. IPCC [15] reports that carbon dioxide capture is the process of reducing carbon dioxide emission from carbon dioxide-producing activities. The main CO_2-producing processes have been identified as industrial- and energy production-related processes.

The capture of CO_2 has aroused a lot of interest among many climate change experts. There are 10 pathways of carbon dioxide capture and sequestration. According to Alden and Hepturn [16] the 10 pathways of CCS and the estimates of the amounts of carbon dioxide each method can remove and the costs associated with each pathway are as follows: (i) CO_2 chemicals—Splitting CO_2 into its constituents, this pathway will take out 0.3–0.6 Gigatonnes of CO2/year in 2050 and will cost from USD 80 to USD 300/tonne of CO_2. (ii) CO_2 fuels—Combining hydrogen and carbon dioxide to form hydrocarbon fuels like methanol, among many others, this pathway will take out 1.0–4.2 Gigatonnes of CO_2/year in 2050 and will cost as much as USD 670/tonne of CO_2. (iii) CO_2 fixation using micro-algae—Algae can be used to fix CO_2, and the resulting biomass is processed into products like fuels and carbon-based chemicals, and this pathway will take out 0.2–0.9 Gigatonnes of CO_2/year in 2050 and will cost from USD 230 to USD 920/tonne of CO_2. (iv) CO_2 use in concrete building materials—This involves using CO_2 to cure concrete works, and this pathway will take out 0.1–1.4 Gigatonnes of CO_2/year in 2050 and will cost from USD 30 to USD 70/tonne of CO_2. (v) CO_2-enhanced oil recovery (EOR)—This involves the injection of CO_2 into oil wells to increase oil production, and this pathway will take out 0.1–1.8 Gigatonnes of CO_2/year in 2050 and will cost from USD 40 to USD 60/tonne of CO_2. (vi) CO_2 capture in bioenergy with carbon capture and storage (BECCS)—This involves splitting CO_2 into its constituents, and this pathway will take out 0.3–0.6 Gigatonnes of CO_2/year in 2050 and will cost from USD 80 to USD 160/tonne of CO_2. (vii) CO_2-assisted weathering—This involves crushing rocks and spreading them over land to accelerate the formation of stable carbonates from CO_2 in the atmosphere, and there are no estimates for this pathway yet. (viii) CO_2 storage by forests—This pathway will take out up to 1.5 Gigatonnes of CO_2/year in 2050 and will cost from USD 10 to USD 40/tonne of CO_2. (ix) CO_2 sequestration by soil—This involves land management for soil carbon management, and this pathway will take out 0.9–1.9 Gigatonnes of CO_2/year in 2050 and will cost from USD 20 to USD 40/tonne of CO_2. (x) CO_2 utilization in agriculture through biochar—This involves utilizing biochar in agriculture, and this pathway will take out 0.2–1.0 Gigatonnes of CO_2/year in 2050 and will cost up to USD 60/tonne of CO_2.

5.2 Transportation of CO_2

A number of researchers have given details on the methods of transport of CO_2. Some reports by [17–19] provide three options for transportation of CO_2 for storage. The options are as follows: (i) CO_2 transportation using pipelines—This transportation option is widely used. Using this method, CO_2 is transported by means of pipelines just like oil and gas. The pipeline can transport CO_2 over very long distances on land. (ii) CO_2 transportation by means of marine tankers—This option of transportation is used after the CO_2 is liquefied. This method is used whenever the distances through which the liquefied CO_2 should be transported far exceed the distances which pipelines can cover. (iii) CO_2 transportation by means of takers and trucks—This method is used to transport liquefied CO_2 by means of trucks and tankers. However, this option is not very attractive [17].

5.3 Storage of CO_2

World Resources Institute [14] and Lal [20], in their studies and subsequent reports, have classified carbon dioxide storage into two broad categories of storage methods. The first method mentioned is the abiotic method. This method does not involve living organisms. Under this method, World Resources Institute [14] and Lal [20] list three categories of procedures as follows: (i) CO_2 storage in oceanic injection—This involves injection of CO_2 streams into the ocean depths. The injection into the ocean depths maybe done by injecting CO_2 below 1000m, injecting CO_2 at 500–1000 m depth as a denser CO_2/seawater mixture, injecting CO_2 as a discharge from a pipe fixed to a ship, or injecting CO_2 by pumping it into the bottom of the sea forming a CO_2 lake. [13] adds onto this by confirming that, actually, oceanic injection of liquid CO_2 at depths of 3000 m and more is known to keep the liquid CO_2 in a stable state for long periods of time. (ii) CO_2 storage by geologic injection—This follows the steps of capture, liquefaction, transportation, and injection of CO_2 into the geologic rocks found in the earth's crust. In this procedure, CO_2 may be injected into coal seams, old oil wells (this increases yield of the oil wells), stable rock formations, or saline aquifers. (iii) Scrubbing and mineral carbonation—This involves transforming industrial CO_2 into $CaCO_3$, $MgCO_3$, and other minerals of similar chemical makeup.

The second method of storage is the biotic method. World Resources Institute [14] and Lal [20] state that this is a method which is based on the managed and/or unmanaged (i.e., natural) intervention in CO_2 storage of plants high up in the plant kingdom (e.g., mangroves) as well as microorganisms (e.g., phytoplankton of the sea and soil among many habitats). World Research Institute [14] and Lal [20] further provide two categories of biotic CO_2 storage as follows: (i) CO_2 sequestration by oceanic organisms—There are a number of biological processes which lead to CO_2 sequestration in the ocean. One such process is the photosynthetic processes carried out by phytoplankton. (ii) CO_2 sequestration by terrestrial processes—This is the transfer of carbon into biotic and pedologic (i.e., soil ecosystems) carbon pools. This is achieved by the action of forest ecosystems (including mangrove forests), wetlands, and the associated soils or histosols.

5.4 Monitoring, measuring, and verification of the impact of the stored carbon dioxide on the environment

The capture, transport, and storage (sequestration) of the carbon should be coupled with monitoring, measuring, and verification (MMV). World Resources Institute [14] accordingly reports that some of the parameters which should be measured, monitored, and verified are (i) the footprint of the project at depth to gather information on CO_2 levels and location and pressure geometry, (ii) pressure and temperature of the reservoirs to gather information on wells' and confining units' integrity, (iii) structural stresses to gather information on the wells' and confining units' integrity, (iv) performance and integrity of the wells to gather information on the wells' integrity evaluation and CO_2 monitoring, and (v) CO_2 concentrations and fluxes to gather information on accidental leakages, accident planning, and early warning systems.

6. Carbon sequestration capabilities of mangroves forests

Greenhouse gases (GHGs) are gases which cause global warming and in turn lead to climate change. Carbon dioxide is the most commonly produced greenhouse gas. The process of carbon sequestration is the process of capturing carbon dioxide from the atmosphere and storing it. The main types carbon sequestration are geologic and biologic [1]. Mangroves play a great role in biologic carbon sequestration. They form ecosystems which scientists refer to as "blue carbon ecosystems" as opposed to "green carbon ecosystems" which are found on the land.

Mangrove forests are able to store three to four times more carbon than the forests which are found on land. Mangroves are able to store and stock pile carbon from the atmosphere during their growing period from 50 metric tons to as much as 220 metric tons per acre. For the whole world, mangroves are therefore able to sequester more than 24 million metric tons of carbon per year [21]. Further, there are other studies showing and confirming that mangroves have great potential in

CO_2 sequestration. One of the supporting studies by Brown et al. [22] reports that geomorphological and biophysical qualities of locations of mangals are important in site selection for maximization of mangrove forest carbon sequestration. On the other hand, Cameron et al. [23] claims that there is not enough scientific evidence due to the small number of studies on the matter. However, despite this, Cameron et al. [23] in their study concluded that aquaculture pond (one of the reasons for mangrove forest destruction) restoration back to mangrove forest ecosystems has great potential for increasing carbon sequestration. Additionally, Spalding et al. [24] reports that if fully restored mangroves forests have the potential to sequester 69,000,000 tonnes of CO_2 which is released into the atmosphere equivalent to a yearly CO_2 atmospheric input by 25 million homes in the United States and save 296,000,000 tonnes of soil carbon stock in avoided emissions which is equivalent to a yearly CO_2 atmospheric input by 117 million homes in the United States.

7. Mangrove destruction

Mangroves have been degraded over time. According to Spalding et al. [24], between the years 1996 and 2016, which is a period of slightly over a decade, total global mangrove forest area had been reduced from 1.42795×10^7 ha to 1.36714×10^7 ha which is equal to a net loss of more than 6.0×10^5 ha. In another study, Muhammed and Koike [25] classified the cases of these losses mentioned by Spalding et al. [24] as either overexploitation or destructive actions caused by activities which do not have any bearing on sustainable exploitation of mangroves.

8. Effects of the destruction of mangrove forests

Mangroves account for 1% of the total biological carbon sequestration. However, they contribute a total of 14% of carbon sequestration by the oceans. This is a huge amount, and it means that if these blue carbon ecosystems are disturbed, the world atmospheric gas system will be upset along with huge amounts of climate change causing carbon dioxide being left in the atmosphere [21].

9. Conclusions

In conclusion, mangroves are plants which occur mostly in the intertidal zone of saline waters. They are special trees which have developed special adaptations to survive the harsh saline environments where they occur. They have huge capabilities for carbon sequestration. This means that these mangrove habitats should be protected as destruction will upset the atmospheric gas balance to a huge extent.

Acknowledgements

The author of this article wishes to thank all great writers who have been cited here. Additionally, the writer wishes to thank the Author Support Manager, Mr Ed Lipovic, for all the support. Further, many thanks is given to Beatrice Obegi for allowing him to use some research photos, and of course, the author does not forget to thank the editor of the book.

Conflict of interest

The author wishes to declare that there is no conflict of interest.

Author details

Charles Nyanga
Zambia College of Agriculture, Zambia

*Address all correspondence to: charles.nyanga@gmail.com

References

[1] UN Environment. Mangroves. [En ligne] 2014. Available from: http://biodiversitya-z.org/content/mangrove-2.pdf [Accessed: 13 February 2020]

[2] American Museum of Natural History. What is a Mangrove? And How Does it Work? [En ligne] American Museum of Natural History, 2020. Available from: http://www.amnh.org/explore/science-bulletins/bio/documentaries/mangroves-the-roots-of-the-sea/what-s-a-mangrove-and-how-does-it-work [Accessed: 12 February 2020]

[3] Coastal Wiki. Mangroves. [En ligne] 2020. Available from: https://coastalwiki.org/wiki/Mangroves [Accessed: 01 January 2020]

[4] Spalding MD, Blasco E, Fielding CD. World Mangroves Atlas. Okinawa: The International Society for Mangrove Ecosystems; 1997

[5] World Wide Fund for Nature. Mangroves Forests: Ecology. Mangroves Forests: Ecology. [En ligne] 2020. Available from: https://wwf.panda.org [Accessed: 02 January 2020]

[6] Ecologic Development Fund. What is a Mangrove? What is a Mangrove? [En ligne] 2019. Available from: https://ecologic.org/actions-issues/about-the-region/What-is-a-mangrove/ [Accessed: 15 November 2019]

[7] Mangroves Action Project. Mangroves Action Project. [En ligne] 2020. Available from: https://mangrovesactionproject.org [Accessed: 20 February 2020]

[8] Wikipedia. Mangroves. Mangroves. [En ligne] 2020. Available from: https://en.wikipedia.org/wiki/Mangroves [Accessed: 03 January 2020]

[9] Reef R, Feller IC, Lovelock CE. Nutrition of mangroves. Tree Physiology. 2010;**30**:1148-1160

[10] American Museum of Natural History. What is a Mangrove? And How does it Work? What is a Mangrove? And How does it Work? [En ligne] 2020. Available from: https://amnh.org/explore/videos/biodiversity/mangrove-the-roor-of-the-sea/what-is-a-mangrove-and-how-does-it-work [Accessed: 20 November 2019]

[11] Bui M, Adjiman CS, Bardow A, et al. Carbon capture and storage (CCS). Carbon Capture and Storage (CCS): The Way Forward. [En ligne] 2018. Available from: http://eprint.whiterose.ac.uk/128986/ ISSN: 17545692. [Accessed: 14 March 2020]

[12] Mather TA. Carbon Capture and Storage (CCS). Cambridge: ESC Publishers, Cambridge University; 2018

[13] Herzbog H. Carbon Dioxide Capture and Storage. [En ligne] 2009. Available from: http://sequatration.mit.edu/pdf/2009_CO2_Capture_and Storage_Ch13_book.pdf [Accessed: 13 March 2020]

[14] World Resources Institute. CCS Guidelines: Guidelines for Carbon Dioxide Capture, Transport and Storage. [En ligne] 2008. Available from: http://pdf.wri.org/ccs_guidelines.pdf [Accessed: 13 March 2020]

[15] Intergovernmental Panel on Climate Change, IPCC. Carbon Dioxide Capture and Storage. Cambridge: Intergovernmental Panel on Climate Change; 2005. p. 443. ISBN-13: 978-0-521-86643-9

[16] Alden E, Hepturn C. 10 Carbon Capture Methods Compared: Costs, Scalability, Performance, Cleanness. 10 Carbon Capture Methods Compared:

Costs, Scalability, Performance, Cleanness. [En ligne] 11 November 2019. Available from: http://energypost.eu/10-Carbon-Captgure-Methods-Compared-Costs-Scalability-Performance-Cleanness [Accessed: 15 March 2020]

[17] Hafez A, Fateen SK. CO_2 transport and storage technologies. In: Sanchez JA, editor. Carbon Dioxide Capture. s.l.: Nova Science Publishers, Inc.; 2016. p. 2

[18] Environmental Protection Agency. Carbon Dioxide Capture, Transport and Storage. [En ligne] 2015. Available from: https://www.epa.gov/sites/production/files/2015-08/documents/chapter_6_carbon_capture_transport_and_storage.pdf [Accessed: 13 March 2020]

[19] Singh U. Carbon capture and storage: An effective way to mitigate global warming. [En ligne] 2013. Available from: https://www.researchgate.net/publication/270159236_Carbon_capture_and_storage_An_effective_way_to_mitigate_global_warming/link/54a1629a0cf256bf8baf6ba9/download [Accessed: 14 March 2020]

[20] Lal R. Carbon Sequestration. [En ligne] 2008. Available from: https://www.researchgate.net/publication/6079761_Carbon_sequestration/link/55420e7b0cf224a89a3333ca/download [Accessed: 14 March 2020]

[21] Twilley R, Rovai, A. Why Protecting 'blue carbon' storage is crucial to fighting climate change. [En ligne] Greenbiz, 2019. Available from: https://greenbiz/article/why-protecting-blue-carbon-storage-is-crucial-to-fighting-climate-change [Accessed: 05 November 2019]

[22] Brown B, et al. Commonly structure dynamics and carbon stock change of rehabilitated mangrove forests in Sulawesi, Indonesia. [En ligne] 2019. Available from: https://www.researchgate.net/publication/330094769_Community_Structure_Dynamics_and_Carbon_Stock_Change_of_Rehabilitated_Mangrove_Forests_in_Sulawesi_Indonesia. DOI: 10.1002/bes2.1478 [Accessed: 14 March 2020]

[23] Cameron C, Hutley BL, Friess DA. Estimating the full greenhouse gas emissions offset potential and profile rehabilitating and established mangroves. [En ligne] 2019. Available from: https://www.researchgate.net/publication/331002073_Estimating_the_full_greenhouse_gas_emissions_offset_potential_and_profile_between_rehabilitating_and_established_mangroves/link/5c6a3e15a6fdcc404eb743ec/download. DOI: 10.1016/j.scitotenv.2019.02.104 [Accessed: 15 March 2020]

[24] Spalding M, Worthinton, et al. Mangrove restoration potential: A global map highlighting a critical opportunity [En ligne] 2019. Available from: https://www.researchgate.net/publication/334141398_Mangrove_restoration_potential_A_global_map_highlighting_a_critical_opportunity [Accessed: 15 March 2020]

[25] Muhammed N, Koike M. Mangrove plantation destruction in Noakhali coastal forests of Bangladesh: A case study on cases consequences and mode prescription to halt deforestation. [En ligne] 2005

Chapter 10

Recycling Polymeric Materials for Corrosion Control

Mohamed A. Deyab

Abstract

The purpose of this chapter is to present the state of the art for using recycled polymeric materials as effective corrosion inhibitors for metals in different corrosive solutions. Initially, the chapter provides the information about corrosion definition, rust formation, and corrosion costs. The chapter gives comprehensive picture about the standard practices for corrosion control, highlighting the recent trends. In addition, it summarizes the corrosion inhibition mechanism in different media. Recent and novel corrosion inhibitors based on recycling several polymeric materials are reviewed and are presented according to the area of application of the inhibitors. As presented, recent trends are focusing on the compounds that are environmentally friendly and less toxic to the environment.

Keywords: recycling polymeric materials, corrosion inhibitors, corrosive solutions, metals

1. Introduction

1.1 Definition of corrosion

Corrosion is a spontaneous process, which transforms a polished metal to a more chemically-stable form (e.g., oxide, hydroxide, or sulfide). It is the gradual destruction of metals by chemical and/or electrochemical reaction with their environment [1–4].

Corrosion has an effect on the microstructure, mechanical features and the physical aspect of the materials. Rusting and other types of deterioration drastically reduce the capacity of pipelines and equipment, resulting in loss of output as well as loss of equipment, or even life.

1.2 Rust formation

Rust is a generic term [5] to describe a series of different oxides, $Fe(OH)_2$, $Fe(OH)_3$, $FeO(OH)$, $Fe_2O_3{\cdot}H_2O$ that form when iron corrodes. The common form of rust is a red product Fe_2O_3 called hematite (**Figure 1**).

1.3 Corrosion costs

According to the recent reports issued by international NACE organization, the global cost of corrosion to be US$2.5 trillion. This cost equivalent to roughly 3.4% of the global gross domestic product (GDP) [6].

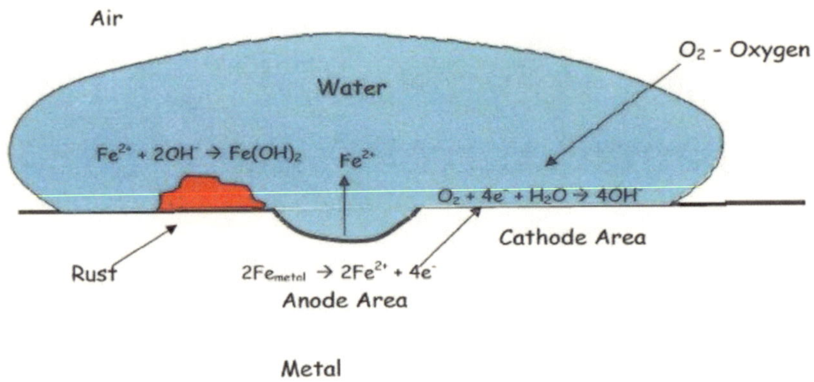

Figure 1.
Rust formation on the metal surface.

1.4 Corrosion control

The corrosion control in different industrial fields can be controlled by the following strategy. They include [7]:

1. Protective coatings

2. Cathodic protection

3. Materials selection

4. Corrosion inhibitors

Protective coatings are effective tools for controlling the corrosion hazards [8–10]. These substances are often applied in conjunction with cathodic protection systems to provide the most cost-effective protection for a structure [11–14].

Cathodic protection (CP) is a technology that uses direct electrical current to stop the corrosion. It is an effective method for underground petroleum storage tank or natural gas pipeline. On new systems, CP can control the corrosion from starting; on existing systems; CP stop the existing corrosion from getting worse.

Materials selection refers to the use of highly corrosion resistant materials such as stainless steels, to enhance the life time of metal systems [15, 16].

Corrosion inhibitors are chemicals compounds that added by small quantities to a particular corrosive media, to overcome on the corrosive hazards of surrounding corrosive media [17–21].

2. Recycling polymeric materials as effective corrosion inhibitors

The recycling of polymeric waste and use of their modified products as effective corrosion inhibitors for metal corrosion are limited [22]. In this section, it is intended to report a series of efforts aimed to the converting plastic waste into useful material used to control the corrosion of metal in very high corrosive environmental.

2.1 Recycling polyethylene terephthalate (PET) plastic waste as corrosion inhibitors

Nowadays, PET plastic show the unique marketing and lifestyle advantages [23–25]. The chemical nature of PET plastic allows to easy recyclability by various recycling methods. Recycling of PET has become an important way from the environmental point of view and it has given commercial opportunity due to wide spread use and availability of PET bottles, packages and fibers [26, 27].

Abdel Hameed et al. [28] prepare some of corrosion inhibitors derived from PET plastic waste. These PET plastic wastes were used to inhibit the corrosion carbon steel in acid corrosive media [29].

Abd El Hameed et al. [22] were prepared new corrosion inhibitors based on the modify (PET), with di-ethanolamine (DEA) and tri-ethanolamine (TEA) using manganese acetate as a catalyst. The prepared corrosion inhibitors were used to protect the carbon steel which used in the manufacture of petroleum pipelines in nitric acid, at different concentrations of inhibitors from 50 to 250 ppm and different temperatures, ranged from 303 to 333 K.

Ayman et al. [30], prepared water soluble corrosion inhibitor from recycled PET. In this work, PET was de-polymerized by TEA into glycolyzed product, followed by esterification with bromoacetic acid in the presence of manganese acetate as a catalyst. The prepared compound was used as corrosion inhibitor for XL65 carbon steel, in 2 M HCl solution. The effectiveness of recycled PET corrosion inhibitor was confirmed by various electrochemical techniques such as open circuit potential, potentiodynamic polarization and electrochemical impedance spectroscopy (EIS). The results of these investigations showed enhancement in inhibition efficiencies with the increasing of inhibitor concentration.

2.2 Recycling poly(ethylene glycol) as corrosion inhibitors

Poly(ethylene glycol) (PEG) is a polyether compound with many applications from industrial manufacturing to medicine [31]. Poly(ethylene glycol) (PEG) is also known as polyethylene oxide (PEO) or polyoxyethylene (POE), depending on its molecular weight. The structure of PEG is commonly expressed as $H—(O—CH_2—CH_2)_n—OH$.

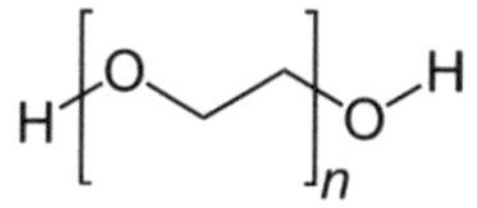

High-molecular weight PEG is synthesized by suspension polymerization of waste materials [32]. It is necessary to hold the growing polymer chain in solution in the course of the poly-condensation process. The reaction is catalyzed by magnesium, aluminum, or calcium-organo-element compounds. To prevent coagulation of polymer chains from solution, chelating additives such as dimethylglyoxime are used.

The use of PEG as corrosion inhibitors was investigated by Deyab and Abd El-Rehim [33]. They studied the effect of PEG of various average molecular weights (MW = 1200, 4000 and 6000) as inhibitors for the corrosion of carbon steel in 1.0 M butyric acid. They found that the corrosion rate of carbon steel in butyric acid decreased with increasing PEG concentration and molecular weight and increased with increasing temperature. PEG inhibits the corrosion of carbon steel by an adsorption mechanism.

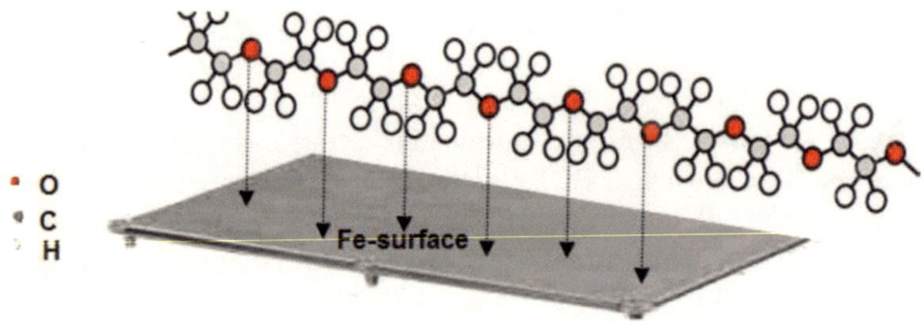

Figure 2.
Showing how a PEG polymer protects the metal from corrosion.

These polymers inhibit the acid dissolution of carbon steel by adsorption at the Fe/acid solution interface. The adsorption process takes place via ion pair and ion exchange mechanism by their ethylene oxide groups [34]. **Figure 2** showing how a PEG polymer protects the metal from corrosion. Inhibition of the PEG can be explained by a substitution adsorption of the polymers according to the following equation.

$$Polymer_{(aq)} + xH_2O_{(ads)} = Polymer_{(ads)} + xH_2O_{(aq)} \quad (1)$$

The data imply that the PEG(6000) has higher corrosion inhibition than PEG(4000) and PEG(1200), indicating that the corrosion inhibition of these polymer increases with the increase in their molecule weights. This is due to the increase in the number of ethylene oxide group [(—CH_2OCH_2—)n] with an increase in the molecular weight of polymer. This causes an increase in the bulkiness of the groups attached to the adsorption center and hence reduces the rate of corrosion [35, 36].

2.3 Recycling copolymer vinyl pyrrolidone/vinyl acetate as corrosion inhibitors

Recycled copolymer vinyl pyrrolidone/vinyl acetate (VP/VA) copolymer was used in many studies as corrosion inhibitors.

In presence of surfactant with polymers in different applications and industrials is greatly noted [37]. The presence of the polymer may be lead to change the properties of the surfactant molecules. The association between polymers and surfactants molecules has been studied by several researchers [38, 39]. Several mechanisms have been proposed to describe this association. These can be summarized as the following:

a. Association of individual surfactant monomers to the polymers chain [40].

b. Association of micelle-like aggregates of the surfactant to the polymer chain [41].

c. A combination group “a” and “b” [42].

The association of polymer-micelle can be detected by [43, 44]:

1. The change in clouding behavior of the polymer,

2. The reduction in Krafft temperature of the surfactant,

3. The decrease in aggregation number,

4. ΔH_{mic} measurements.

In the present part, we report recent studies upon the association between surfactant and recycled VP/VA copolymer and its effect on the adsorption properties of the surfactant molecules on metal surfaces. The study review consists of two major parts: the association between anionic (AS) and cationic (CS) surfactant and recycled VP/VA copolymer and the effect of the association on the adsorption properties of the surfactant molecules on the corrosion of carbon steel in acidic solution.

The conductometric measurements were used to investigate the critical micelle concentration (CMC) of ionic surfactants. The break point in the plot of conductivity versus surfactant concentration point to the CMC values of the surfactant [45].

The addition of VP/VA copolymer molecules shifts the CMC values of surfactant to lower value. Many literature data [46, 47] reveal that the reduction in CMC, points to polymer-micelle association. One of the most convincing models for polymer-surfactant association has always been the binding of surfactant molecules (S) to the polymer molecules (P) through a stepwise sequence of several chemical equilibrium [48]:

$$PS_{n-1} + S \underset{K_{-n}}{\overset{K_n}{\rightleftarrows}} PS_n \tag{2}$$

Polymer-surfactant association in solution depends on the balance forces between charge-charge interactions between surfactant head group and the dipoles of a non-ionic polymer; the hydrophobic interaction arises from water molecules and the dispersion forces between nearest-neighbor molecules [49]. On other hand, the data reveal that the presence of polymer molecule does not exert any significant influence on CMC values of cationic surfactant CS. This behavior is consistent with the fact there is no significant association comparable to CS surfactant with neutral copolymer. Many literature data suggest that the association between the surfactant and polymer are probably due to a partially positive charge transfer to the CO group of the polymer and, consequently, an increased attraction of the hydrophilic group of the anionic surfactants [49, 50]. This postulate would explain the absence or presence of very weak association between cationic surfactant and neutral copolymer.

Various factors directly influence on cationic surfactant-neutral copolymer association such as solution pH (see **Figure 3**) and additions of KCl (see **Figure 4**) to test solutions are discussed herein.

The data clearly show that in pH range for 3.4–4 there is no significant changes in CMC values of CS surfactant in the absence and presence of copolymer. On the other hand, an increase in the solution pH to 6 and 8 values will significantly increase the CMC of the CS surfactant as a result of increasing the association process. This behavior may be attributable to the increase of solution pH (in range 6–8) which leading to a reduction of the positivity of CS surfactant and, consequently, an increased attraction of the hydrophilic group of the cationic surfactant to the copolymer.

It is apparent that in the absence of KCl there are no significant changes in CMC values of CS surfactant in the absence and presence of copolymer. On other hand, addition of different KCl concentrations to CS surfactant solutions in the absence and presence of neutral copolymer decreases the CMC of the CS surfactant. This result suggests that the presence of KCl in the solutions enhances the association between CS surfactant and copolymer molecules. The extent of association increases with increasing KCl concentration form 0.02 to 0.06 mM. This behavior is due mainly to a screening effect. The clusters or pre micelles of CS surfactant which are adsorbed on the copolymer chain are surrounded by counter ions (Cl^-). As a consequence the electrical repulsion between CS surfactant and copolymer molecules is decreased and association takes place at a lower concentration of surfactant [51, 52].

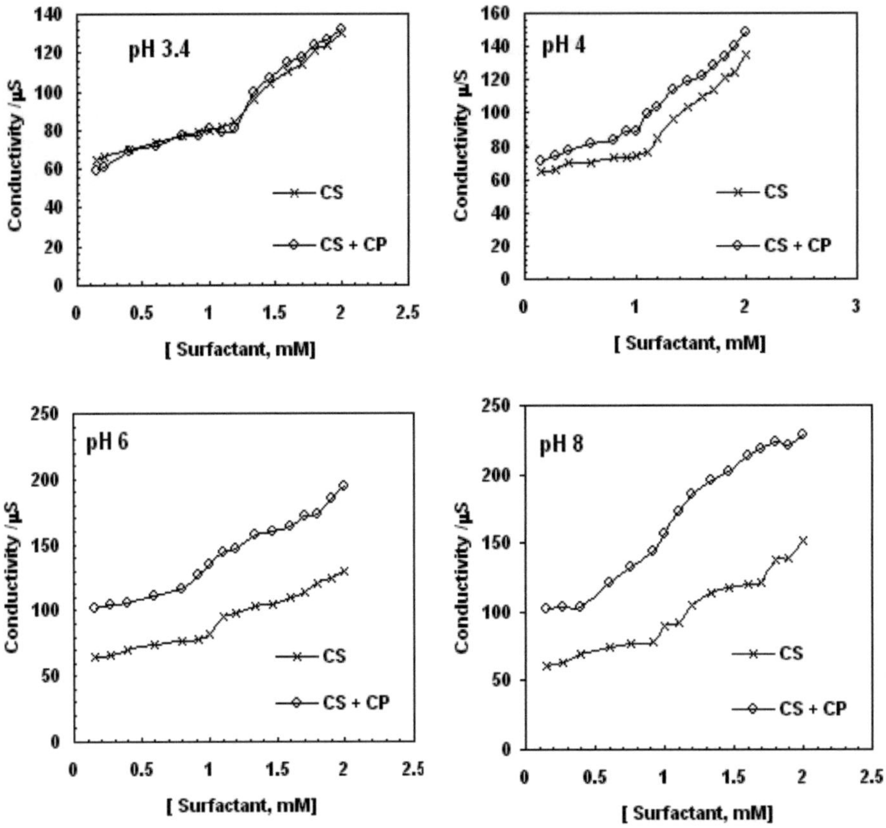

Figure 3.
Effect of solution pH on CMC values of CS surfactant.

Form the corrosion studied, it was found that the corrosion protection efficiency (P%) for the CS surfactant in the absence and presence of copolymer increased with increase in the concentration of CS until it reaches a maximum constant value corresponding to the critical micelle concentration (CMC = 0.8 mM) [53]. The protective behavior of CS may be due to adsorption of surfactant on the metal surface. The presence of 0.05 gm/l of polymer has no significant changes in protection efficiency P% and CMC values of CS surfactant. This may be suggesting that the association between CS surfactant and neutral copolymer under this condition is very week. Consequently, the corrosion inhibition of metal is mainly due to the adsorption of CS surfactant on the metal surface and formation of barrier film that separating the metal surface from the corrosive acid medium.

The corrosion protection efficiency of CS surfactant in the absence and presence of copolymer increases with increasing the pH values. Here we also observed that the effect of pH of solution on the protection efficiency of CS surfactant. This is due to the presence of copolymer enhanced the inhibition effect of CS especially in pH range 6–8. This can be explained on the basis that the partial positive charge of copolymer decreases in pH range 6–8 and, consequently, an increased attraction to the hydrophilic group of the cationic surfactant. The obtained explanation is compared with that obtained by above conductivity studies. According to this behavior, the association between CS and copolymer molecules leads to a higher degree of coverage and this, in turn, leads to an increase in protection efficiency P% at a given pH value.

The protection efficiency of CS in the absence of copolymer increases with increasing KCl concentration. This behavior can be interpreted on the basis that addition of KCl to the solution enhances the negative charge on the metal surface and this leads to

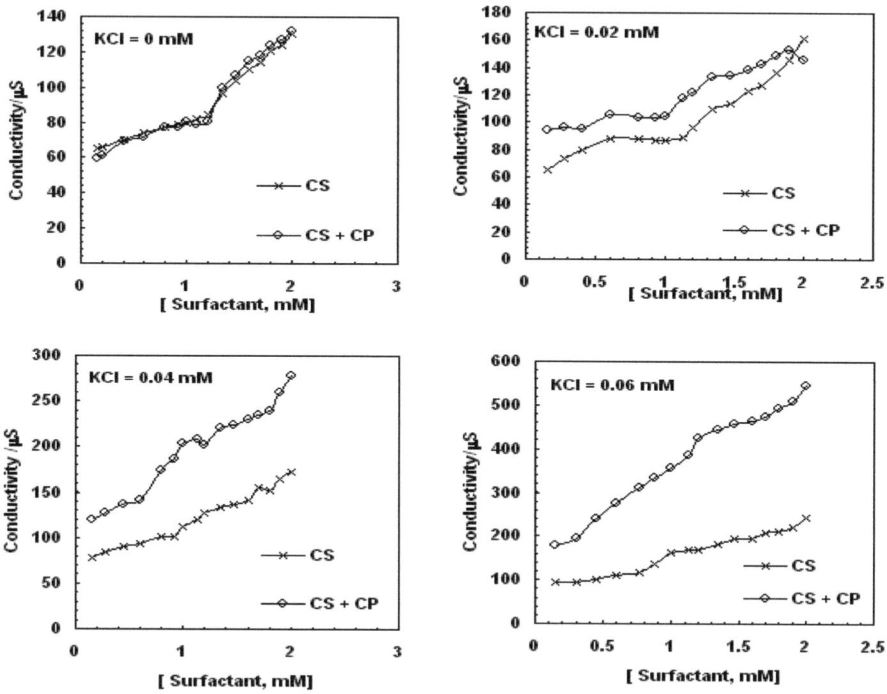

Figure 4.
Effect of KCl concentration on CMC values of CS surfactant.

an increase in adsorption of CS on the metal surface and in turn, leads to an increase in protection efficiency P% [54]. The presence of polymer molecules with Cl^- ions and CS is accompanied by high protection efficiency as compared with corrosive medium contains Cl^- ions and CS only. It is clear that the change of the protection efficiency (P_{CS+CP}%–P_{CS}%) increase with increasing the KCl concentration. This observation can be explained on the basis that the presence of KCl increase the adsorption of CS on the metal surface as a result of increasing the negative charge on the metal surface as well as the enhancing of the association between CS and copolymer molecules because of decreasing the electrical repulsion between CS and copolymer molecules [54]. This leads to greater surface coverage, thereby giving higher inhibition efficiency.

Different types of adsorption may be considered for the adsorption of AS surfactant molecules at the metal surface: (i) electrostatic attraction between charged molecules and the charged metal, (ii) interaction of unshared electron pairs in the molecule with the metal, and (iii) a combination of the above [55, 56].

The presence of VP/VA polymer with AS surfactant increase protection efficiency of AS and this behavior may be due to polymer-surfactant association leading to a higher degree of coverage and consequently higher corrosion inhibition [57]. The protection efficiency for AS in the absence and presence of polymer increases with increase in AS concentration until it reaches a maximum constant value corresponding to its critical micelle concentration. The maximum protection efficiency of AS in the absence of polymer (65%) was achieved at 0.67 mol/dm^3. On other hand the presence of polymer enhances maximum protection efficiency of AS (73%) and achieves it at lower concentration.

2.4 Recycling polystyrene (PS) as corrosion inhibitor

A survey of pervious works revealed the using of polystyrene (PS) as corrosion inhibitor in acidic media.

Shittu et al. [58] indicated in his study that an addition of PS in small quantity reduced the corrosion rate of steel in acidic solutions (HCl and H_2SO_4). Optical inspection of the steel surface under testing showed that there was formation of thin layer on the metal surface and the layer thickness increases as the polystyrene concentration increases. This PS layer isolates the steel surface from the corrosive solution.

Románszki et al. [59] they used polystyrene layers thickness (85–500 nm) as effective coating on copper and copper alloys as well as on stainless steel 304 to control the corrosion in artificial seawater.

2.5 Recycling polyurethane (PU) as corrosion inhibitor

Kumar et al. [60] had reported polyurethane tri-block co-polymer as effective inhibitor of mild steel corrosion in acidic medium. The data indicated that the PU inhibits metal corrosion by adsorbing on metal surface to form pseudo-capacitive interface. Scanning electron microscopy (SEM) and atomic force microscopy (AFM) analyses showed that the polymers formed protective film on MS surface and shield it from direct acid attack.

Banerjee et al. [61] investigated a series of polyurethane (PU) with different degree of sulfonation (DS) as corrosion inhibitor for mild steel in acidic solution. More than 90% inhibition efficiency polyurethanes (PU) have been reported using only 20 ppm of the PU. Polyurethane (PU) inhibited the corrosion of steel through adsorption process. Surface coverage by PU has been scaled using scanning electron microscopy (SEM) and atomic force microscopy (AFM).

3. Conclusion

The recycling of polymeric waste and use of their modified products as effective corrosion inhibitors for metal corrosion are limited. In this chapter, it is intended to report a series of efforts aimed to the converting plastic waste into useful material used to control the corrosion of metal in very high corrosive environmental. Many of the reports deal with novel polymeric corrosion inhibitors in various applications were discussed in this chapter.

Author details

Mohamed A. Deyab
Petroleum Research Institute (EPRI), Cairo, Egypt

*Address all correspondence to: hamadadeiab@yahoo.com

References

[1] Putilova I, Balezin S, Barannik V. In: Ryback G, translator. Metallic Corrosion Inhibitors. New York: Pergamon Press; 1960

[2] Uhlig HH. Corrosion and Corrosion Control. 2nd ed. New York: John Wiley & Sons; 1971

[3] Boffardi BP. Corrosion inhibitors in the water treatment industry. In: ASM Handbook. Vol. 13A. Materials Park, OH: ASM International; 2003

[4] Shreir LL, Jarman RA, Burnstein GT. Principles of Corrosion and Oxidation, Corrosion. 2nd ed. Elsevier; 1994

[5] Landrum RJ. Fundamentals of Designing for Corrosion Control. Houston, TX, USA: NACE International; 1992

[6] Roberge PR. Corrosion Engineering Principles and Practice. New York: McGraw-Hill; 2008

[7] Speller FN. Corrosion Causes and Prevention. 3rd ed. New York: MeGraw-Hill; 1951

[8] Deyab MA. Effect of carbon nano-tubes on the corrosion resistance of alkydcoating immersed in sodium chloride solution. Progress in Organic Coatings. 2015;**85**:146-150

[9] Deyab MA, Keera ST. Effect of nano-TiO_2 particles size on the corrosion resistance of alkyd coating. Materials Chemistry and Physics. 2014;**146**:406-411

[10] Deyab MA. Corrosion protection of aluminum bipolar plates with polyaniline coating containing carbon nanotubes in acidic medium inside the polymer electrolyte membrane fuel cell. Journal of Power Sources. 2014;**268**:50-55

[11] Deyab MA, Eddahaoui K, Essehli R, Benmokhtar S, Rhadfi T, Riccardis AD, et al. Influence of newly synthesized titanium phosphates on the corrosion protection properties of alkyd coating. Journal of Molecular Liquids. 2016;**216**:699-703

[12] Deyab MA, Riccardis AD, Mele G. Novel epoxy/metal phthalocyanines nanocomposite coatings for corrosion protection of carbon steel. Journal of Molecular Liquids. 2016;**220**:513-517

[13] Deyab MA, Nada AA, Hamdy A. Comparative study on the corrosion and mechanical properties of nano-composite coatings incorporated with TiO_2 nano-particles, TiO_2 nano-tubes and ZnO nano-flowers. Progress in Organic Coatings. 2017;**105**:245-251

[14] Deyab MA, Mele G, Al-Sabagh AM, Bloise E, Lomonaco D, Mazzetto SE, et al. Synthesis and characteristics of alkyd resin/M-porphyrins nanocomposite for corrosion protection application. Progress in Organic Coatings. 2017;**105**:286-290

[15] Deyab MA, Keera ST. Cyclic voltammetric studies of carbon steel corrosion in chloride-formation water solution and effect of some inorganic salts. Egyptian Journal of Petroleum. 2012;**21**:31-36

[16] Deyab MA. The influence of different variables on the electrochemical behavior of mild steel in circulating cooling water containing aggressive anionic species. Journal of Solid State Electrochemistry. 2009;**13**:1737-1742

[17] Clayton CR, Olefjord I, Marcus P, Oudar J. Corrosion Mechanisms in Theory and Practice. New York: Marcel Dekker; 1995

[18] Antropov LI, Makushin EM, Panasenko VF. Inhibitors of Metal Corrosion. Moscow: Metallurgia; 1976

[19] Deyab MA, Keera ST, El Sabag SM. Chlorhexidine digluconate as corrosion inhibitor for carbon steel dissolution in emulsified diesel fuel. Corrosion Science Journal. 2011;**53**:2592-2597

[20] Deyab MA. Effect of halides ions on H_2 production during aluminum corrosion in formic acid and using some inorganic inhibitors to control hydrogen evolution. Journal of Power Sources. 2013;**242**:86-90

[21] Deyab MA. Hydrogen generation during the corrosion of carbon steel in crotonic acid and using some organic surfactants to control hydrogen evolution. International Journal of Hydrogen Energy. 2013;**38**:13519-13511

[22] Abd El Hameed RS, Shafey HIAL, Soliman SA, Metwally MS. Corrosion of C-steel alloy in (0.1M) Nitric acid in the presence of plastic waste as corrosion inhibitors. Al-Azhar Bulletin of Science. 2008;**19**:283

[23] Colomines G, Robin J, Tersac G. Study of the glycolysis of PET by oligoesters. Polymer. 2005;**46**:3230-3247

[24] Al-Salem SM, Lettieri P, Baeyens J. Recycling and recovery routes of plastic solid waste (PSW): A review. Waste Management. 2009;**29**:2625-2643

[25] Chilton T, Burnley S, Nesaratnam S. A life cycle assessment of the closed-loop recycling and thermal recovery of post-consumer PET. Resources, Conservation and Recycling. 2010;**54**:1241-1249

[26] Dullius J, Ruecker C, Oliveira V, Ligabue R, Einloft S. Chemical recycling of post-consumer PET: Alkyd resins synthesis. Progress in Organic Coatings. 2006;**57**:123-127

[27] Achilias DS, Roupakias C, Megalokonomos P, Lappas AA, Antonakou EV. Chemical recycling of plastic wastes made from polyethylene (LDPE and HDPE) and polypropylene (PP). Journal of Hazardous Materials. 2007;**149**:536-542

[28] Abdel Hameed RS, Al-shafey HI, Ismail EA. Studies on corrosion inhibition of C- steel in 1 M acetic acid solutions by ethoxylated poly(ethylene terphthalate) derived from plastic waste. Al-Azhar Bulletin of Science. 2009;**20**:185-197

[29] Kim J, Park S, Moon I. Corrosion control document database system in refinery industry. Chemical Engineering. 2009;**27**:1839-1844

[30] Ayman MA, Shehata HA, Abd El Bary HM, Abdel Salam S, Abdel Hameed RS. Recycled poly(ethylene terephthalate) waste oligomers as corrosion inhibitors of steel. Progress in Rubber, Plastics and Recycling Technology. 2007;**23**:209-226

[31] Kahovec J, Fox RB, Hatada K. Nomenclature of regular single-strand organic polymers. Pure and Applied Chemistry. 2002;**74**:1921-1956

[32] Chua BW, Lee CS, Lim WH, Pichika MR. One-pot synthesis of cobalt-incorporated polyglycerol ester as an antimicrobial agent for polyurethane coatings. Journal of Applied Polymer Science. 2018;**135**:46045

[33] Deyab MA, Abd El-Rehim SS. Influence of polyethylene glycols on the corrosion inhibition of carbon steel in butyric acid solution: Weight loss, EIS and theoretical studies. International Journal of Electrochemical Science. 2013;**8**:12613-12627

[34] Deyab MA. Hydrogen evolution inhibition by L-serine at the negative electrode of a lead-acid battery. RSC Advances. 2015;**5**:41365-41371

[35] Deyab MA. Inhibition activity of seaweed extract for mild carbon steel corrosion in saline formation water. Desalination. 2016;**384**:60-67

[36] Deyab MA, Essehli R, El Bali B. Inhibition of copper corrosion in cooling seawater under flowing conditions by novel pyrophosphate. RSC Advances. 2015;**5**:64326-64334

[37] Nagarajan R. Association of nonionic polymers with micelles, bilayers, and microemulsions. The Journal of Chemical Physics. 1989;**90**:1980

[38] Goddard ED. Polymer-surfactant interaction. Part I. Uncharged water-soluble polymers and charged surfactants. Colloids and Surfaces. 1986;**19**:255

[39] Schich MJ. Nonionic Surfactants. New York: Marcel Dekker; 1967

[40] Lewis KE, Robinson CP. The interaction of sodium dodecyl sulfate with methyl cellulose and polyvinyl alcohol. Journal of Colloid and Interface Science. 1970;**32**:539

[41] Minatti E, Zanette D. Salt effects on the interaction of poly(ethylene oxide) and sodium dodecyl sulfate measured by conductivity. Colloids and Surfaces A: Physicochemical and Engineering Aspects. 1996;**113**:237

[42] Schwuger MJ. Mechanism of interaction between ionic surfactants and polyglycol ethers in water. Journal of Colloid and Interface Science. 1973;**43**:491

[43] Goddard ED. Interactions of Surfactants with Polymers and Proteins. Boca Raton: CRC Press; 1993

[44] Rizzatti IM, Zanette DR, Zanette D. Construction of surfactant-membrane electrodes selective for sodium dodecyl sulfate in poly(ethylene oxide)-surfactant mixtures. Journal of the Brazilian Chemical Society. 2004;**15**:491

[45] Jones MN. The interaction of sodium dodecyl sulfate with polyethylene oxide. Journal of Colloid and Interface Science. 1967;**23**:36

[46] Zana R, Lianos P, Lang J. Fluorescence probe studies of the interactions between poly(oxyethylene) and surfactant micelles and microemulsion droplets in aqueous solutions. The Journal of Physical Chemistry. 1985;**89**:41

[47] Chari K. Polymer-surfactant assembly at an interface. Journal de Physique II France. 1995;**5**:1421

[48] Gharibi H, Rafati AA, Feizollahi A, Razavizadeh BM, Safarpour MA. Thermodynamic studies of interaction between cationic surfactants and polyvinyl pyrrolidone using potentiometric techniques. Colloids and Surfaces A: Physicochemical and Engineering Aspects. 1998;**145**:47

[49] Yépez O. Influence of different sulfur compounds on corrosion due to naphthenic acid. Fuel. 2005;**84**:97

[50] Keera ST, Deyab MA. Effect of some organic surfactants on the electrochemical behaviour of carbon steel in formation water. Colloids and Surfaces A: Physicochemical and Engineering Aspects. 2005;**266**:129

[51] Dubin PL, Gruber JH, Xia J, Zhang HJ. The effect of cations on the interaction between dodecylsulfate micelles and poly(ethyleneoxide). Colloid and Interface Science. 1992;**148**:35

[52] Xia J, Dubin PL, Kim Y. Study of polymers in equilibrium with random obstacles. The Journal of Physical Chemistry. 1992;**96**:6805

[53] Atia AA, Saleh MM. Inhibition of acid corrosion of steel using

cetylpyridinium chloride. Journal of Applied Electrochemistry. 2003;**33**:171

[54] Deyab MA. Effect of cationic surfactant and inorganic anions on the electrochemical behavior of carbon steel in formation water. Corrosion Science. 2007;**49**:2315

[55] Solmaz R, Kardaş G, Yazıcı B, Erbil M. Adsorption and corrosion inhibitive properties of 2-amino-5-mercapto-1,3,4-thiadiazole on mild steel in hydrochloric acid media. Colloids and Surfaces A: Physicochemical and Engineering Aspects. 2008;**312**:7

[56] Kutej P, Vosta J, Pancir J, Hackerman N. Electrochemical and quantum chemical study of propargyl alcohol adsorption on iron. Journal of the Electrochemical Society. 1995;**142**:1847

[57] Emregül K, Atakol O. Corrosion inhibition of iron in 1 M HCl solution with Schiff base compounds and derivatives. Materials Chemistry and Physics. 2004;**83**:3730

[58] Shittu MD, Olawale JO, Adeoye MO, Oluwasegun KM, Adebayo KM, Ige OO. Investigation of corrosion resistance of polystyrene as an inhibitor in hydrochloric and tetra-oxo sulphate VI acids. International Journal of Materials and Chemistry. 2014;**4**:9

[59] Románszki L, Datsenko I, May Z, Telegdi J, Nyikos L, Sand W. Polystyrene films as barrier layers for corrosion protection of copper and copper alloys. Bioelectrochemistry. 2014;**97**:7

[60] Kumar S, Vashisht H, Olasunkanmi LO, Bahadur I, Verma H, Singh G, et al. Experimental and theoretical studies on inhibition of mild steel corrosion by some synthesized polyurethane tri-block co-polymers. Scientific Reports. 2016;**6**:30937

[61] Banerjee S, Mishra A, Singh MM, Maiti B, Ray B, Maiti P. Highly efficient polyurethane ionomer corrosion inhibitor: The effect of chain structure. RSC Advances. 2011;**1**:199-210